Anouk Majerus

Influence of climatic factors on the radial growth of junipers

Anouk Majerus

Influence of climatic factors on the radial growth of junipers

at an alpine dry site

Imprint

Any brand names and product names mentioned in this book are subject to trademark, brand or patent protection and are trademarks or registered trademarks of their respective holders. The use of brand names, product names, common names, trade names, product descriptions etc. even without a particular marking in this work is in no way to be construed to mean that such names may be regarded as unrestricted in respect of trademark and brand protection legislation and could thus be used by anyone.

Cover image: www.ingimage.com

This book is a translation from the original published under ISBN 978-620-0-44971-9.

Publisher:
Sciencia Scripts
is a trademark of
Dodo Books Indian Ocean Ltd. and OmniScriptum S.R.L publishing group

120 High Road, East Finchley, London, N2 9ED, United Kingdom
Str. Armeneasca 28/1, office 1, Chisinau MD-2012, Republic of Moldova, Europe
Printed at: see last page
ISBN: 978-620-4-21703-1

Contents

"Never limit yourself because of the limited ingenuity of others; never limit others because of your own limited ingenuity."
-Jane Goodall

Acknowledgement

First of all, I would like to thank Prof Dr Walter Oberhuber and Prof Dr Georg Wohlfahrt. They supported and advised me during this time and helped me a lot. I learnt a lot from them.

I would also like to thank my dear friends Astrid Steger, Lisa Jung, Claudia ..., Krishna Magnussen, Manuel Eberl, Anna Rottensteiner and Sarah Oberleiter. They always had an open ear for me, supported me, showed me new perspectives and even proofread.

A very special thanks goes to my parents, Danielle Schmitz Majerus and Henri Majerus, who have always supported me and always had an open ear for me. During these years they had a great influence on my life and without them I would not be where I am today. My father also helped me with the core sampling and travelled with me to Mieming.

I would also like to thank my godmother Claude Kartheiser and my brother Laurent Majerus. They were a great support and valuable listeners, a true enrichment of my life.

Summary

Periods of drought in spring restrict the radial stem growth of conifers, which dominate in dry inner-alpine valleys. In the dry, sparse pine forest on the Mieminger Plateau (Tyrol, 910-950 metres above sea level), the common juniper (*Juniperus communis*) can be found in the undergrowth. *Juniperus communis* is considered to be very drought-resistant and undemanding of nutrients. Occasionally, tree-shaped specimens with trunk heights >2 m have developed, which were investigated in this study using dendroocological methods. Three hypotheses were formulated. HI: The radial growth of *Juniperus communis* is limited by water availability, i.e. low precipitation in spring and high temperatures during the growing season. H2: Due to the increase in evapotranspiration, climate warming leads to a decrease in basal area increment (BFE), especially in individuals with low vitality. H3: *Juniperus communis* shows a higher drought resistance in radial growth compared to *Pinus sylvestris*. 120 increment cores were taken from *Juniperus communis* at a stem height of 30 cm from individuals of different vigour. In addition, 15 increment cores were analysed from the *Pinus* sylvestris *stand*. The assignment to the three vitality classes (VK) was carried out on the basis of crown thinning (KV). The annual ring width (ANW) was measured with a resolution of 1 pm and the growth chronologies were checked for correct dating of the annual increments. The GFZ, the extreme growth years and the stress indices (resistance, recovery and resilience) were calculated. It was found that the radial growth of *Juniperus communis* has been increasingly limited by precipitation in April-June in recent decades and that the GFZ has risen sharply in the course of management changes (reduction or cessation of grazing), but has reached a plateau since around 2014 due to increasing competition from *Pinus sylvestris*. However, the *Juniperus* communis *individuals* with low vitality recorded a slump in their growth, while the individuals of medium vitality showed a constant and those of high vitality an increasing GFZ. There were extreme declines in growth in 1993, 1997, 2005 and 2023. The GFC of *Pinus sylvestris* remained constant in the period from 1900 to 2023, while that of *Juniperus communis* showed a significant increase between 1982 and 2014, presumably due to changes in management, but this was offset by the increasing population density in the last few years.

reached a plateau in the second decade. The better adaptation of *Pinus sylvestris* to the local site factors is shown by the fact that the SPC of *Juniperus communis* only reaches about 1/3 of the SPC of *Pinus sylvestris*. Ecophysiological surveys are necessary to clarify in detail the cause of the conspicuously lower growth of *Juniperus communis* compared to *Pinus sylvestris*.

1. introduction

1.1 Dendrology and dendrochronology

Dendrology is concerned with the study of trees and woody plants and thus represents an important area within botany. Within botany, dendrology is specifically assigned to woody plants, shrubs, trees and climbers (Bartels, 1993).

Dendrochronology deals with the temporal assignment of tree ring width variations (Buntgen et al., 2005). It is a method of dating in which the width of the annual rings is assigned to a specific period of tree growth (Buntgen et al., 2006). Annual rings from years with favourable growth conditions are wider than those from years with unfavourable conditions (Buntgen et al., 2005). In young years, trees normally form wider annual rings, which then gradually become narrower (Buntgen et al., 2006). However, this is only possible in regions with clear periods of vegetation and dormancy due to the changing activity of the cell division tissue (cambium).

The growing season typically begins in spring, which also marks the start of cell division activity in the cambium (Sheppard, 2010). In conifers, wide-lumened and dark-walled cells known as earlywood are formed at this time (Sheppard, 2010). As the growing season progresses, cell division in the cambium decreases (Sheppard, 2010). Narrow lumened, dark-coloured and thick-walled cells are then formed, known as spar wood (Sheppard, 2010). These cells mainly have a function for the stability of the tree trunk.

The growth period ends in autumn and does not begin again until the following spring. The alternation between earlywood and latewood creates a clearly recognisable boundary between the annual rings, which makes it possible to determine the age of the tree (Sheppard, 2010).

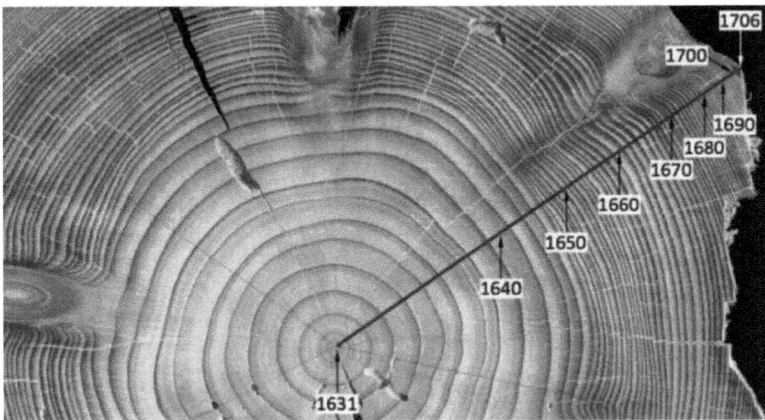

Figure 1: Cross-section of a tree trunk. Growth rings visible from 1631 to 1706 ©Gottfried Letschke
https://www.ecology.uni-Jena.de/80/dendro-oekologie

1.2 Dendroocology and dendroclimatology

Dendroocology investigates the environmental factors that influence the radial growth of trees and shrubs (Sheppard, 2010; Schweingruber, 1996). Changes in land use over time can also influence growth (Cook et al., 2015). In this way, insights can be gained into the response of woody plants to environmental factors and predictions can be made for growth behaviour under the expected conditions of climate change (Anchukaitis, 2017).

The growth of trees depends on many environmental conditions, including water availability, sunlight, temperature, nutrients, site dynamics, pest infestation and forestry measures (Bohm, 2010).

As climatic factors can influence thickness growth, analysing annual rings also allows conclusions to be drawn about climatic conditions. If the JRB of many individuals at a location decreases sharply in the same year, this indicates an extreme year. In this way, information can be obtained about environmental influences such as climate, disturbance factors etc. that have led to this reduction in growth. This is possible by superimposing the pattern of tree rings of several individuals to obtain averaged growth time series (tree ring chronologies) (Cooketal., 2015).

Dendroclimatology is concerned with reconstructing the influence of climate on tree ring growth (Wilson et al., 2016). However, the challenge is to filter out the correct climate signals (Sheppard, 2010). Other factors can override the climate signals, such as damage to the tree, competition from other trees, wind pressure and earth movements (Anchukaitis, 2017).

1.3 Goals& Hypotheses

In dry inner-alpine valleys, periods of drought in spring restrict the growth of trees and shrubs (Pichler & Oberhuber, 2007; Ellenberg & Leuschner, 2010; Drexler, 2020). In recent decades, there has been an increase in tree mortality, which is accelerated by these drought periods (IPCC, 2021). In the sparse pine forest on the Mieminger Plateau, where isolated specimens of the common juniper (*Juniperus communis*) with a tree-shaped habitus occur, the effect of climatic factors and climate warming on the radial stem growth of tree-shaped individuals is being investigated. The results are compared with the growth time series of *Pinus sylvestris*. The sensitivity to drought stress (stress indices) and the long-term trend in the GFZ as a function of the vitality (crown thinning) of tree-shaped *Juniperus* communis *individuals* will also be focussed on. The results should provide information on the climate sensitivity of *Juniperus communis*.

Three hypotheses were developed. HI: The radial growth of *Juniperus communis* is limited by water availability, i.e. low precipitation in spring and high temperatures during the growing season. H2: Due to the increase in evapotranspiration, climate warming leads to a decrease in basal area increment (BFE), especially in individuals with low vitality. H3: *Juniperus communis* shows a higher drought resistance in radial growth compared to *Pinus sylvestris*.

2 Material and methods

2.1 *Juniperuscommunis*

The common juniper (*Juniperus communis*) is known by many names, such as heath juniper, Machandel tree, Kranewitt tree, Reckolder, incense tree or fire tree. *Juniperus communis* belongs to the Cupressaceae family (cypress family) (Schutt et al., 2007).

Figure 2: Tree-shaped individual of Juniperus communis.

Juniperus communis is versatile in its growth forms, either upright like a tree or creeping along the ground. *Juniperus communis* can reach a height of up to 12 metres, and it has been documented that the tallest *Juniperus communis* ever recorded was 18.5 metres high, with an impressive trunk diameter of 0.9 metres (Roloff etal. 2007). *Juniperus communis* can reach a remarkable age of up to 600 years. The bark is typically grey to reddish-brown in colour, while the crown is usually narrow and conical to oval in shape (Roloff et al. 2007).

The leaves of *Juniperus communis* are needle-shaped and sit on the twig through a joint, arranged in whorls of three needles each. These needles are pointed and measure about 1-2 centimetres in length. A striking feature is the light-coloured stripe on the upper side of the needles, which has stomata and wax streaks (Roloff et al. 2007).

Juniperus communis is a dioecious species, although monocious specimens occasionally occur (Jagel, 2012). Male plants are easily recognisable by their yellowish flowers during the flowering period from April to June. The cones are formed in autumn and are provided with a stalk (Jagel, 2012). Female bleeding cones consist of three cone scales in which the ovules are located. Over time, the seed scales fuse with the covering scales and become fleshy. Maturation into berry-shaped cones takes about three years (Jagel, 2012).

Juniperus communis, a deep-rooter with mycorrhiza, spreads its pollen with the help of the wind during its flowering season from April to May. As one of the most widespread conifers, its habitat extends from North America to South Africa, North Africa, Europe, the Near East, North Asia, Central Asia and East Asia, up to an altitude of 4,000 metres (Roloff & Bartels, 2006).

Although *Juniperus communis* is a rather uncompetitive plant and is often crowded out by other species, it can be quite dominant on dry, sandy, stony sites or moorland. It thrives particularly well on sunny rough pastures, on rocks and in sparse woodland. This species prefers dry, sunny and mostly base-rich, calcareous soils (Roloff & Bartels, 2006) and is considered an indicator of former grazing (Ellenberg and Leuschner, 2010).

2.2 *Pinus sylvestris*

Pinus sylvestris, also known as Scots pine or Scots pine, is an important tree species in Europe and Asia (Weber et al., 2007). It belongs to the pine family (Pinaceae) and is one of the most common tree species in European forests (Schmidt, 2003). The Scots pine is an evergreen conifer that can reach heights of up to 30 metres and forms a conical crown. The needles of the pine are arranged in pairs and are about 4-7 cm long. The bark is typically fissured and often reddish-brown in colour.

This tree species is extremely adaptable and thrives in a wide range of habitats, from dry sandy soil to moist moorland (Wimmer & Kriebitzsch, 2002). However, it prefers locations with plenty of sunlight and is tolerant of drought and frost (Schutt and Stimm, 2006). The cones of the Scots pine contain seeds that are dispersed by the wind (Weber et al., 2007).

2.3 Study area

The study area is located on the Mieminger Plateau and comprises an Erico-Pinetum forest in which *Juniperus communis* is widespread in the shrub layer.

Figure 3: Erico-Pinetum

The study area is located at an altitude of 910 to 950 metres at the foot of the Mieminger Gebirge. A forest research station of the Institute of Ecology of the University of Innsbruck (FAIR Forest Station Mieming) is located in the pine stand, whose geographical coordinates are 47° 18.9957'N; 10° 58.2016'0 (Fluxnet Austria, 2024). Around this station, 120 cores were taken from tree-form *Juniperus* communis *individuals* in a radius around the forest

station, the study area is shown in Figure 4. The south-facing slope has a low inclination (<10
).°

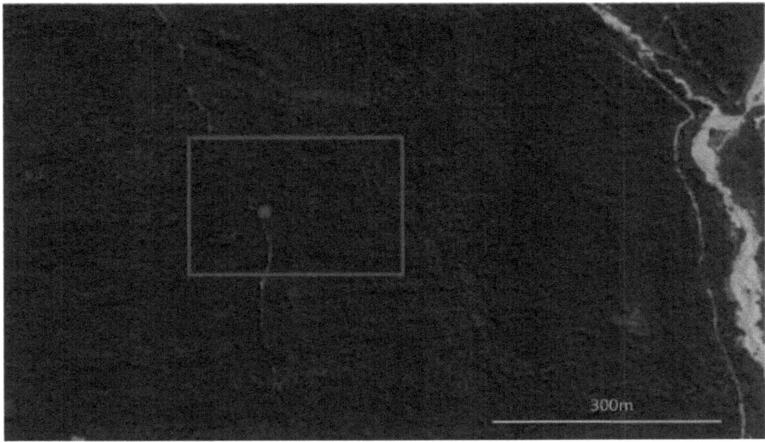

Figure4: Aerial photograph of the 2023 study area (https://earth.google.com/)
Red dot: Mieming Fair Station The red rectangle contains the study area

The Mieminger Gebirge is a mountain range in North Tyrol and is bordered by the
Wetterstein mountains in the north, the Inntal valley in the south, the Fernpass in the west
and the Leutasch valley in the east. Almost all of the rocks in the Mieminger Kette were
formed on the seabed and consist of limestone and its transformation product, dolomite.
However, there are also sandstones, mudstones, cherts, greywackes and volcanic tuffs
(Mandi G. W., 1996; Baumgarten B., 2014).

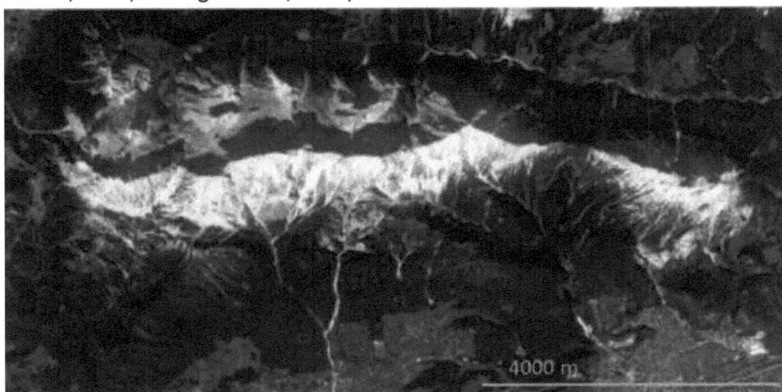

Figure 5: Mieminger Gebirge (https://earth.google.com/); (red rectangle: study area)

The soil of the study area is calcareous with a low humus layer and a low water storage
capacity. The inner alpine dry valley is characterised by low precipitation with an average
annual precipitation of 776 mm and an average annual temperature of 9.2° C. The low
humus cover together with the low precipitation makes the site a dry location and exposes
the vegetation to drought stress. These conditions have led to the establishment of the

9

drought-tolerant *Pinus sylvestris* and *Juniperus communis*. In the tirisMaps, the predominant forest stand is labelled as dry carbonate pine forest (Kl2c) (Figure 7) and lies on gravel limestone (KiK; see Figure 6).

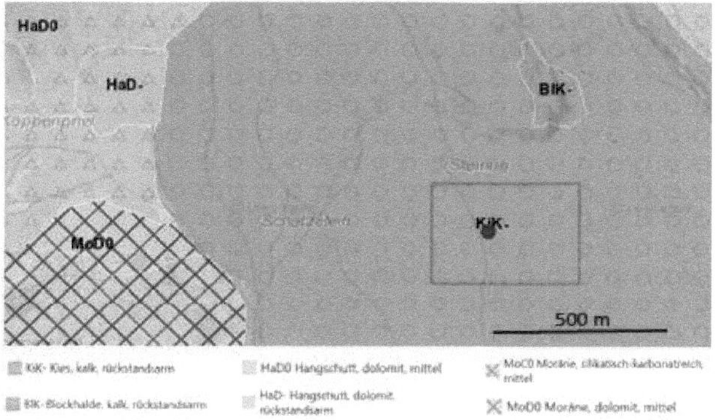

Figure 6: Information on the soil/substrate of the forest sites category
Green rectangle: research area; red circle: forest research station with the coordinates 47°13.9957'N; 10°58.2016'Q Source: tirisMaps, Land Tirol BEV

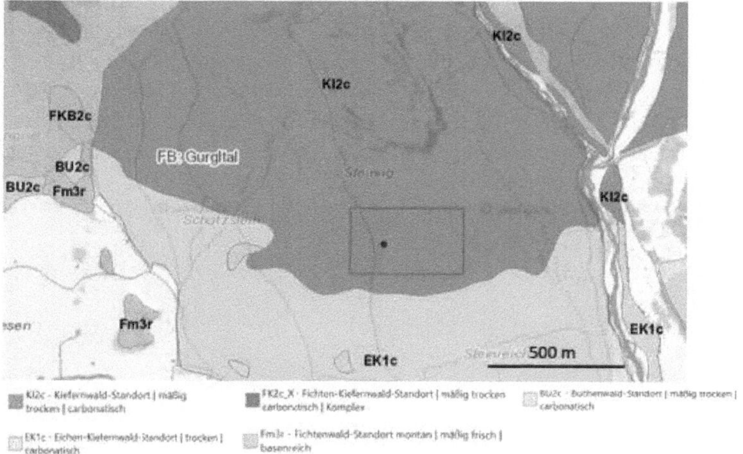

Figure 7: Information on the forest types in the forest sites category
green rectangle: study area; red circle: forest research station with the coordinates 47°13.9957'N; 10°58.2016'Q
Source: tirisMaps, Province of Tyrol BEV

The prevailing conditions favour the growth of *Pinus sylvestris* and *Juniperus communis*, as both species are drought-tolerant. *Pinus sylvestris*, like *Juniperus communis*, is a light plant and has a broad ecological spectrum that enables it to thrive in both very moist and very dry habitats. Due to its prickly needles, this evergreen shrub also spreads as a "pasture weed" in sparse pine stands (Ellenberg and Leuschner, 2010).

According to the Tyrolean spatial information system tirisMaps (https://mapsmobile.drol.gv.at/), the forest on the study area is classified as a protection forest. For this reason, the forest is not used for economic purposes, but primarily serves as a recreational area. There are hiking and cycling trails as well as a hunting ground within the area.

Figure 8 shows the precipitation and temperature curve for the months of April to June for the period 1951 to 2023 (climate station Otz; approx. 13.5 km from the study area as the crow flies). Precipitation remains fairly constant over the years, but some dry periods can be observed. Particularly outstanding is the year 2004 with only 94.3 mm of precipitation, with an average of 200 mm. You can clearly see that the temperature rises significantly over the years. The April to June temperature is currently 13.1 °C (average for the years 2000 to 2023; MW 11.9 °C from 1951 to 2023).

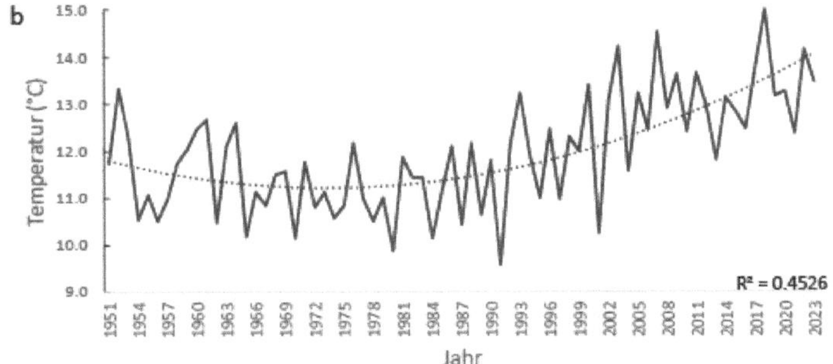

Figure 8: Precipitation and temperature curve from April to June from 1951 to 2023
(a) Precipitation (b) Temperature (trend lines polynomial function; N: y = 0.0162x^2 - 0.8385x + 210.15; T: y= 0.0012x^2 - 0.0547X+11.854)

2.4 Drill core extraction and data collection

Tree-shaped individuals with a minimum height of 1 m and a trunk circumference of > 7 cm were selected to take the cores from *Juniperus communis*. The samples were taken at the beginning of October, at a time when radial stem growth had already been completed for the year. An increment borer (Mattson, 40 cm long) was used to take the cores. Unlike usual, the sample was not taken at breast height, but at ground level (approx. 30 cm stem height). The drill was inserted horizontally into the trunk until the tip of the drill broke through the bark at the other end. The sample was then removed, labelled and placed in a plastic tube. Clamping forceps held the drill core in the increment borer while it was pulled out of the tree-shaped individual.

The total diameter could be taken from each tree-shaped individual. Each individual was also categorised into a vitality class (VK). This was done on the basis of crown thinning (CT), whereby less than 10 % was regarded as high vitality, 10-50 % as medium vitality and less than 50 % as low vitality. In the end, 49 individuals were categorised as high VC, 36 as medium VC and 33 as low VC.

The coordinates, soil depth, trunk circumference and trunk height were also measured at each location. The coordinates were determined using a mobile phone and Google Maps. The soil depth was measured by imprinting a clamping tongue up to the rock. The trunk circumference was measured with a measuring tape and the trunk height with a measuring stick

measured.

Figure 9: Drill and drill core name ©Haglof Sweden https://haglofsweden.com/

The removed drill cores were air-dried to prevent fungal infestation. The drill cores were then attached to a square bar with a groove using a hot glue gun. The fibre direction of the drill core was aligned vertically to ensure correct measurement of the JRB cross-section. The drill cores were cut with industrial blades in order to clearly recognise and measure the individual growth rings in the cross-section. Chalk was then rubbed onto the smooth surface to make the growth rings and their boundaries even more clearly visible by allowing the chalk dust to penetrate the lumina.

Of the 120 cores taken, 14 cores could not be analysed due to breakage or growth anomalies.

In addition, 15 cores of *Pinus sylvestris* were taken from the same study area to supplement the growth time series compiled by Drexler (2020). The cores were taken at breast height (1.3 m) from dominant old trees.

2.5 Measurement of the growth ring width

The growth rings were measured using a movable linear measuring system (LinTab4, Rinntech) and a reflected light microscope (Olympus SZ61). In the right eyepiece of the microscope there was a crosshair that was placed at the annual ring boundary of the next medullary year. The distance was measured from the inside to the outside by moving the measuring system. Ideally, the measurement process begins at the first growth ring boundary. At each annual ring boundary, a foot switch was pressed to send the distance to the computer and save it. It was important to ensure that the crosshairs were always perpendicular to the growth ring boundary in order to avoid measurement errors. Measurements were taken with an accuracy of 1/1000 mm. The variable JRB resulted in different growth time series, also known as chronologies. As most of the cores contained two radii, a mean chronology was created from the two growth chronologies for each tree-form individual.

2.6 Dating and creation of growth chronologies

The *Pinus* sylvestris *chronology* (n=100 individuals) compiled by Drexler (2020) was used as a reference chronology. As no reference chronology was available for *Juniperus communis* and *Pinus sylvestris* was studied at the same site, this growth chronology was a suitable comparison. When selecting the reference time series, however, it was not primarily the growth chronology itself that was taken into account, but rather the extreme growth years. This was due to the fact that two different plant species were involved. If the extreme years matched between the two species, it could be assumed that the measurements and dating were correct. Missing growth rings could have led to incorrect dating.

COFECHA (version 6.06P) was used to correct errors in the dating of the tree ring chronologies created in order to compare each individual tree ring series with a master chronology. Of the 106 growth time series of *Juniperus communis* created, 26 had to be excluded for further analysis due to individual growth fluctuations and/or missing growth rings. This left a sample number of 80 tree-form *Juniperus communis* individuals, which were distributed among the VC 1 (n= 36), 2 (n= 25) and 3 (n= 19).

The synchronisation of the tree ring series was carried out using the TSAP Win program, whereby the parameters overlap length (OVL), similarity (GLK), t-value and statistical significance were determined. The overlap length (OVL) describes the period over which two chronologies overlap. This value indicates how many years or annual rings two samples have in common, which provides important information for the temporal classification and synchronisation of tree ring data (Fritts, 1976). The similarity (GLK) indicates the similarity of two tree-ring chronologies and is given as a percentage. A statistically significant GLK indicates an identical growth response of the compared time series to environmental factors (Holmes, 1983). This measure is crucial in order to

This allows conclusions to be drawn about environmental conditions and growth conditions that influence tree growth. The TVBP value (t-value according to Baillie and Pilcher, 1973) is a measure of the synchronisation of two chronologies. A TVBP value of at least 3.5 indicates a high degree of synchronisation between the chronologies. Statistical significance is defined by the p-value. A p-value below 0.05 indicates that the correlation is significant and the probability of a random correlation is low. As the p-value decreases, the significance of the correlation increases.

2.7 Increase in land area

The GFZ describes the annual increase in the number of stems. One advantage of the GFZ compared to the JRB is that there is no long-term "age trend" in the GFZ as there is in the JRB. This means that long-term growth trends can be analysed.

The GFZ was calculated using the following formula:

$$GFZ = \pi \left(r_a^2 - r_i^2 \right)$$

Here, r_a stands for the radius of the annual ring boundary and n for the radius of the inner ring. The GFZ was calculated based on the 80 mean time series or individuals.

A one-sided t-test for independent samples was used to determine the differences in the GER for the period from 1976 to 2016 between the CVs (class 1, class 2 and class 3).

2.8 Cambial age

When visualising radial growth by cambial age, the JRBs or GFZs are displayed using the annual ring sequence (starting with 1 for the first annual ring in the medullary zone). This makes it possible to identify major or age-related trends in radial growth. 80 *Juniperus communis individuals* were divided into two age groups: < 50 years and > 50 years. The age group < 50 years accounted for 34 individuals and the age group > 50 years for 46 individuals. In addition, the individuals were divided into three classes according to their vitality. A total of 36 individuals were assigned to high CV, 25 individuals to medium CV and 19 individuals to low CV.

2.9 Determination of extreme years

The GFZ was used to determine extreme growth years. The formula for calculating the extreme value is

$$Extreme\ growth\ year = \frac{GFZ_x \times \bar{x}(GFZ_{3x})}{STD}$$

GFZ_x = GFZimJahrx

$\bar{x}$ = Mean value

GFZsx = GFZ of the three previous years of year x

STD = Standard deviation in the GFZ of all years

This formula was used to calculate the extreme growth years for the period from 1911 to 2023.

2.10 Standardisation and the climate-growth relationship

The term age trend stands for all age-related changes in annual increment and their course, speed and the influence of internal and external factors. The age trend is thus defined as the decrease in JRB with increasing age and stem height due to the increase in the ever-widening stem circumference (Schweingruber, 1983).

In dendrochronology, standardisation, also known as indexing, enables JRBs from different drill cores to be compared with each other. Long-term biological-occological fluctuations, which make a direct comparison difficult, are eliminated. Ecological changes in environmental conditions, such as changes in competition and endogenous influences such as age and size trends, influence the radial growth of plants and lead to long-term growth fluctuations (Fritz, 1976; Biasing et al., 1984; Gruber et al, 2010). Indexing makes it possible to draw conclusions about climatically induced changes in growth.

With the help of the programme ARSTAN Version 6.02 (Holmes 1994), indexing is made possible for both single chronologies and mean chronologies. ARSTAN uses a twofold trend elimination procedure with the aid of autoregressive modelling (residual chronologies). The measured value of the current year (Xi) is divided by the measured value of the equalisation function *(xf)* (Schweingruber, 1983).

$$I_i = \frac{x_i}{x_i^0}$$

I_t = Index value

x_i = Measured value of the measured value series

x_i^0 = Measured value of the equalisation function (trend value of the respective year)

The index chronology only shows the deviations from the long-term, standardised mean, which means the value fluctuates around 1. The following statistical parameters were collected. The mean sensitivity is a MaR that expresses in per cent how strong the change between two consecutive values in a time series is. It refers to the mean value of the chronology and shows how strongly the JRB fluctuate on average from year to year (Schweingruber, 1983). The autocorrelation is a measure of the similarity within a time series with a temporal shift of one to several years. It provides information about the correlation of a chronology. If the previous year has a strong influence on the following year, then the correlation coefficient is significant. Here the

time series are shifted by one tree ring and then compared or synchronised (Schweingruber 1983). The standard deviation is a measure of the dispersion of the values from the calculated mean value. The signal-to-noise ratio (SNR) provides information on the homogeneity of the radial growth of the sampled individuals, with a high value indicating high data quality (generally a pronounced climate signal). The Expressed Population Signal (EPS) is a measure of the consistency of tree ring data within a sample. A value close to 1 indicates high representativeness and agreement of the data (Wigley et al., 1984). The explained variance (EV 1) indicates the intensity of the influence of climatic factors on radial growth (Fritts, 1976).

2.11   Moving climate-growth correlation

The DendroClim 2002 program (Biondi, 1997; Biondi and Waikul, 2004) was used to analyse the statistical relationship between tree growth and climate with regard to their temporal changes. Monthly climate variables (temperature and precipitation) and indexed tree ring series were used as starting values. The analysis was carried out step by step for both single and multiple time intervals. The moving response function method works with a fixed number of years (period) that is shifted stepwise to calculate the coefficients (Biondi, 1997;

Biondi and Waikul, 2004). The calculations were carried out over an overlapping period of 40 years, starting in 1911 and ending in 2023. Temperature averages and precipitation totals from September of the previous year to August of the current year were used for the temperature and precipitation data. The precipitation and temperature data are from the Otz weather station from 1911 to 2023. The temperature data are averaged for each month and the precipitation totals are totalled per month.

2.12 Stress indices

The stability of an ecosystem against stress factors can be defined by three parameters (stress indices): Resistance, recovery and resilience. These parameters are used to analyse the effects of drought stress on the three SCs. According to Lloret et al. (2011), they are defined as follows:

Resistance: Resistance describes the extent to which performance is reduced during a disruption. It is determined by the ratio between the power during and before a disruption. Resistance values can be greater than, less than or equal to 1. Values below 1 indicate that growth during the drought period is below the growth performance before the drought phase.

Recovery: Recovery describes the ability to recover after being affected by a disturbance event. Recovery values below 1 indicate a decline in growth after a dry phase.

Resilience: Resilience refers to the ability to regain the ability to perform before a disruptive event. It is determined by the ratio between performance before and after a disruption. If the resilience falls below the value of 1, the growth performance after the drought period is lower than before this phase.

These parameters are calculated using the GFZ.

$$\text{Resistance} = \frac{GFZ_x}{\bar{x}\,(GFZ\ Vorjahr_{3x})}$$

$$\text{Recreation} = \frac{\bar{x}\,(GFZ\ Folgejahr_{3x})}{GFZ_x}$$

$$\text{Resilience} = \frac{\bar{x}\,(GFZ\ Folgejahr_{3x})}{\bar{x}\,(GFZ\ Vorjahr_{3x})}$$

GFZ_x = GFZimJahrx

GFZ previous year x = GFZ of the three previous years of year x

GFZ following yearx = GFZ of the three following years of yearx

$\bar{x}$ = Mean value

3 Results

3.1 Tree age, trunk diameter, trunk height and soil depth

Table 1 shows the statistical parameters of the tree-ring chronologies of the *Juniperus communis* stand (n = 80). The oldest individual is 126 years old and the youngest individual is 16 years old. The mean age a\\of *Juniperus communis* individuals is 51.9±22.7 years and the mean JRB is 0.979±0.189 mm. The low autocorrelation value (AC; r = 0.1404) indicates a low influence of the previous year on the growth of the following year. MS (0.2) shows that *Juniperus communis* reacts sensitively to environmental conditions. Thus, environmental conditions have an influence on radial stem growth. The SNR of 10.05 indicates a high data quality. The EPS of 0.971 indicates a good representation of the sample of a population. The EVI is 24.08 %.

Table 1 also shows the statistical parameters of all test classes (SCs). VC 1 (n = 36) has a mean age of 45±21.9 years. The maximum age is 126 years and the minimum age is 16 years. The mean JRB is 0.98910.121 mm and the AC is 0.0349. The MS is 0.14, which indicates a lower sensitivity compared to the total population. The SNR at 6.763 indicates good data quality. The EPS has a value of 0.958 and the EVI is 26.6%, which shows a strong dominance of the main trend in the data.

UK 2 (n = 25) has a mean age of 31129.7 years. The maximum age is 126 years and the minimum is 17 years. The mean JRB is 0.97710.21 mm and the AC is 0.208. The MS value is 0.193, which shows a higher sensitivity compared to VC 1. The SNR value of 2.408 indicates a lower data quality. The EPS is 0.898, indicating a slightly lower representation of the population, and the EVI is 26.61%.

UK 3 (n = 19) has a mean age of 28121.9 years. The maximum age is 109 years and the minimum age is 21 years. The mean JRB is 0.97710.27 mm and the AC is 0.0652. The MS value is 0.267, which shows the highest sensitivity among the CVs. The SNR at 2.316 indicates the lowest data quality. The EPS value is 0.912 and the EVI is 23.98 %.

Table 1: Statistical parameters of the chronologies of Juniperus communis
n = number; JRB = tree ring width; AC = autocorrelation; MS = mean sensitivity; SNR = signal-to-noise ratio; EPS = expressed
population signal; EVI = explained variance (calculated using residual chronologies; 1898 - 2023)

	n	Age (years) Min/Max	Age (years) MW±STD	JRB (mm) MW±STD	AC	MS (%)	SNR	EPS	EVI (%)
Juco_80	80	16/126	51.9±22.7	0.979±0.189	0.1404	0.2	10.05	0.971	24.08
Vit 1	36	16/126	45±21.9	0.989±0.121	0.0349	0.14	6.763	0.958	26.6
Vit2	25	17/126	31±29.7	0.977±0.21	0.208	0.193	2.408	0.898	26.61
Vit3	19	21/109	28±21.9	0.977±0.27	0.0652	0.267	2.316	0.912	23.98

As can be seen in Figure 10a, *Juniperus communis* *individuals* are categorised by age group as follows: One individual is under 20 years old, 55 individuals are in the age group of 21 to 55 years old, 42 individuals are between 56 and 90 years old, 6 individuals are between 91 and 125 years old, and 2 individuals are 126 years old. The average age of the sampled individuals at the height of the core sampling is 58.5 years. 40 individuals have a diameter of less than 4.5 cm, 32 individuals have a diameter between 4.6 and 5.5 cm, 19 individuals have a diameter between 5.6 and 6.5 cm, 16 individuals have a diameter between 6.6 and 7.5 cm, and 13 individuals have a diameter of more than 7.6 cm (see 10b). The mean diameter is 5.4

cm. The stem height distribution in Figure 10c shows that most individuals have a height between 1 and 3 metres. Specifically, 42 individuals are less than 275 cm tall, 36 individuals are between 276 and 350 cm tall, 16 individuals are between 351 and 425 cm tall, 12 individuals are between 426 and 500 metres tall and 14 individuals are over 500 cm tall. The mean value here is 338 cm. Figure 10d illustrates the considerable variation in soil depth. This was measured for each individual. A soil depth of less than 4 cm was recorded at 4 study sites, at 22 sites it was between 4.1 and 6 cm, at 37 sites between 6.1 and 8 cm, at 28 sites between 8.1 and 10 cm and at 29 sites the soil depth was over 10 cm. The average soil depth was 8.5 cm.

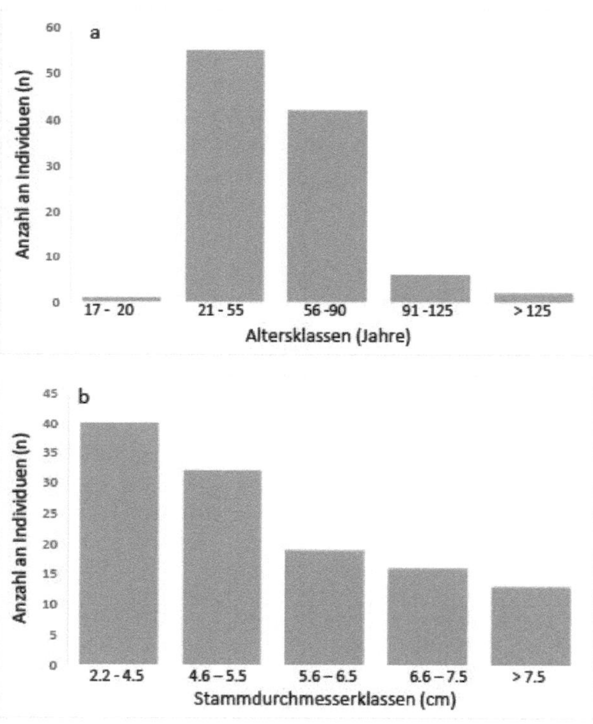

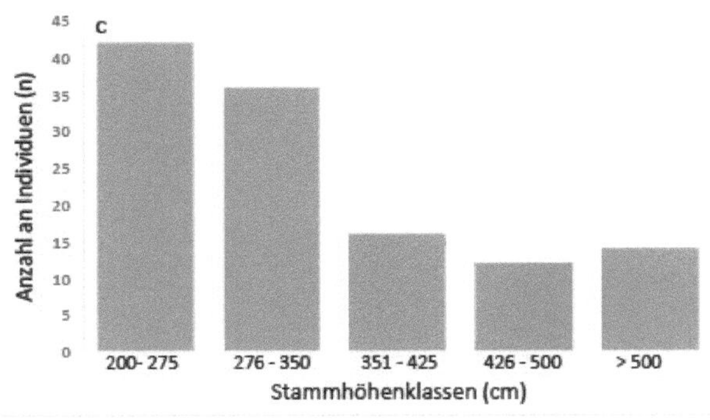

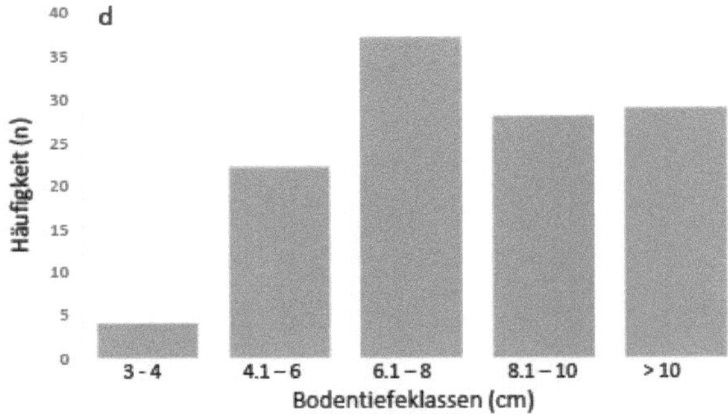

Figure 10: Frequency distribution of age, trunk diameter, trunk height and soil depth
(a) age classes (b) stem diameter classes (c) stem height classes (d) soil depth classes (n = 106)

The trunk diameter and trunk height show a low correlation (P>0.05), with R^2 having a value of 0.1016 (see Figure 11a). Figure lib shows that the soil depth and stem height do not correlate with each other. The correlation between soil depth and stem diameter shows an R^2 value of 0.0395, which indicates that there is no correlation between the two parameters here either (see Figure 11c). Figure lid shows no correlation between age and stem height. There is no correlation between age and stem diameter (R^2 = 0.0589; see Figure lie).

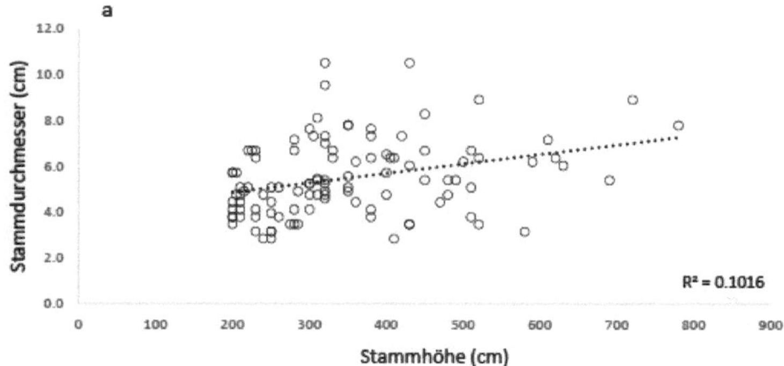
a
Stammdurchmesser (cm)
12.0
10.0
8.0
6.0
4.0
2.0
0.0
0
100
200
300
400
500
600
700
800
900
Stammhöhe (cm)
R² = 0.1016

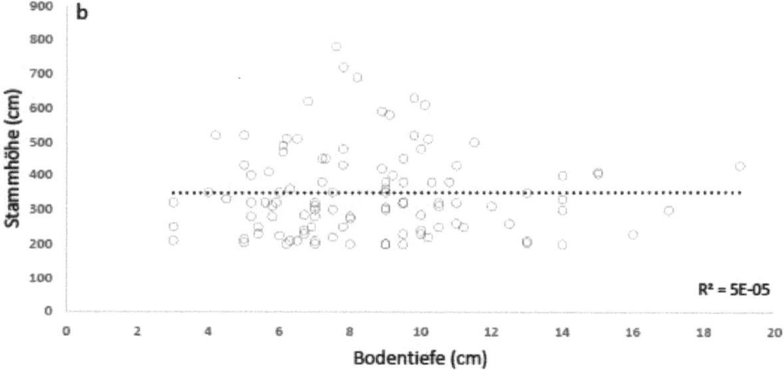
b
Stammhöhe (cm)
900
800
700
600
500
400
300
200
100
0
0
2
4
6
8
10
12
14
16
18
20
Bodentiefe (cm)
R² = 5E-05

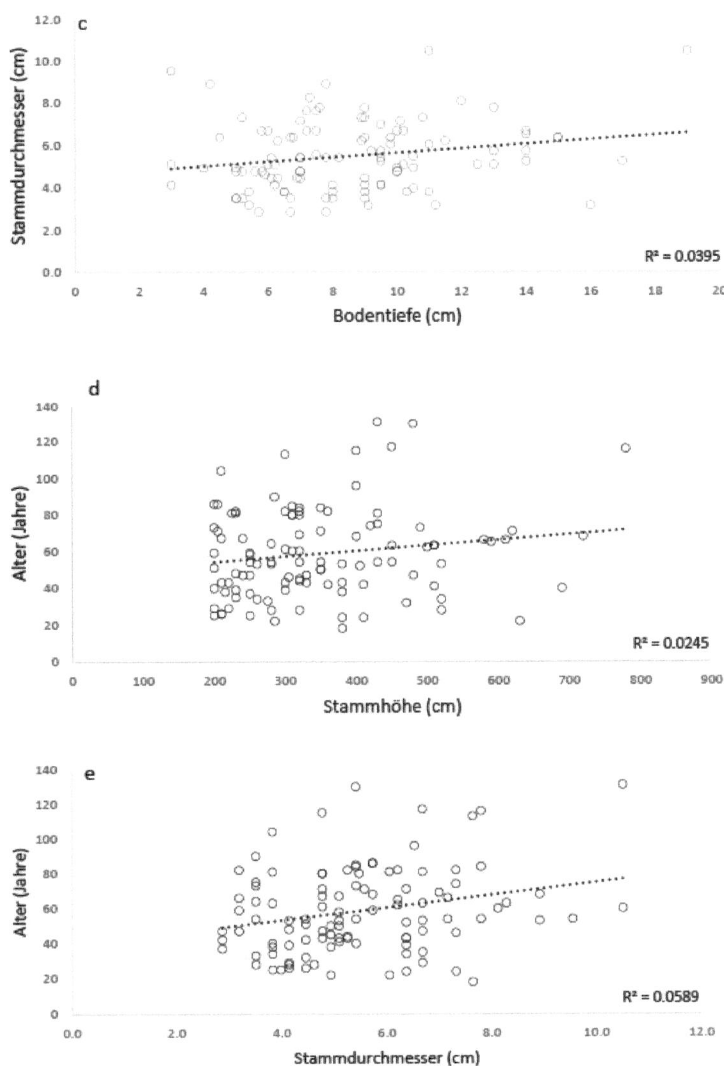

Abbildung 11:   Linear correlation between: (a) stem diameter and stem height (b) soil depth and stem height (c) soil depth and stem diameter (d) age and stem height (e) age and stem diameter   (n = 106)

Figure 12 illustrates the soil depth in relation to the recorded CV of *Juniperus communis*. It shows that the decrease in vitality of the individuals is related to the soil depth. The mean soil depth is 9.4 ± STD cm for individuals of SC 1, 8.3 ± STD cm for SC 2 and 7.5 ± STD cm for SC 3.

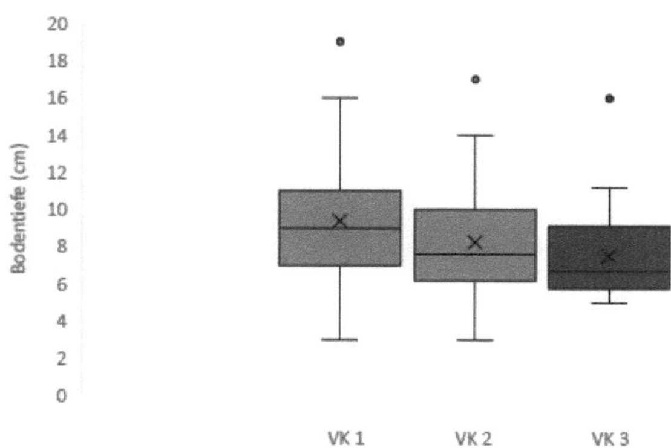

Abbildung 12:   Soil depth in relation to the vitality of the sampled individuals
Each box plot represents the soil depth of the respective vitality. The upper point shows the deepest measured soil depth per vitality. (Whiskers = standard deviation; line in boxplot = median value; x = mean value; green = VK 1 (n = 36); orange = VK2(n = 25); red = VK3 (n = 19))

12.2   Statistical parameters of the growth time series

The mean chronology of 80 individuals (2410080_mean) and that of VK 1 (24101_Vit), 2 (24102_Vit) and 3 (24103_Vit) were compared with the pine chronology (K241_MKn78; Drexler 2020) (Table 2).

Table 2: Statistical values of the given tree ring chronologies with the pine chronology K241_MKn78 by Drexler (2020)
2410080_Mittel stands for the mean chronology of 80 synchronous Juniperus communis.
24101_Vit stands for the middle chronology with the Juniperus communis of VK 1, 24102_Vit stands for VK2 and 24103_Vit stands for VK3. (OVL = overlap length, GLK = synchronisation, TVBP = t-value Baillie and Pilcher)

Chronology	OVL	GLK	TVBP
2410080_Mediu m	121	54	1.4
24101_Vit	121	56	3
24102_Vit	106	54	1.5
24103_Vit	105	45	0.7

Table 3 shows the comparison of the three UKs with the statistical values. The comparison of UK 1 and UK 2 shows a similarity of 69%, while the comparison between UK 2 and UK 3 and also between the time series UK 1 and UK 3 shows a similarity of 66%. A TVBP value above 4.5 indicates a high degree of synchronisation between two chronologies. It is 5.2 for UK 1 and UK 2, 2.2 for UK 2 and UK 3 and 1.6 for UK 1 and UK 3. This shows that UK 1 and 2 are highly synchronised. VK 2 and VK 3 have significantly poorer synchronisation and VK 1 and VK 3 the lowest.

Table 3: Statistical values and comparison of the three UKs
VKI(n = 36); VK2(n = 25); VK3(n = 19)
*OVL = overlap length, GLK = synchronisation, TVBP = t-value Baillie and Pilcher, DateL = start date, DateR = end date, *** = P<0.001*

Sample	Ref.	OVL	GLK		TVBP	DateL	DateR
VK1	VK2	111	69	***	5.2	1898	2023
VK2	VK3	110	66	***	2.2	1913	2023
VK1	VK3	110	66	***	1.6	1898	2023

12.3 Growth chronologies

Figure 13 shows the JRB chronologies of all 80 *Juniperus* communis *individuals* (grey) over the period from 1898 to 2023. The averaged time series shows a slight increase in the period shown (1898 to 2023). The maximum increase is 2.73 mm and the minimum increase is 0.028 mm. With the increase in the number of samples, the dispersion of the JRB chronologies also increases.

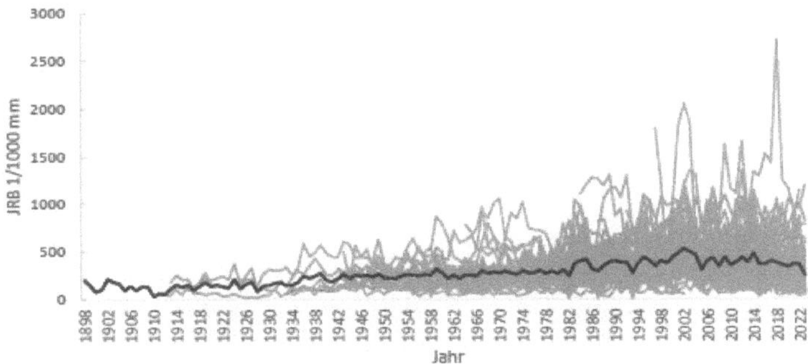

Figure 13: Tree-ring width chronologies (JRB) of all dated Juniperus communis individuals (n=80) and their mean chronology (red line).

Figure 14 shows the mean JRB of 80 synchronised chronologies of *Juniperus communis* for the period from 1898 to 2023. A sample size of 5 is given from 1935 onwards. A slight increase of approximately 150 pm can be seen over the entire period.

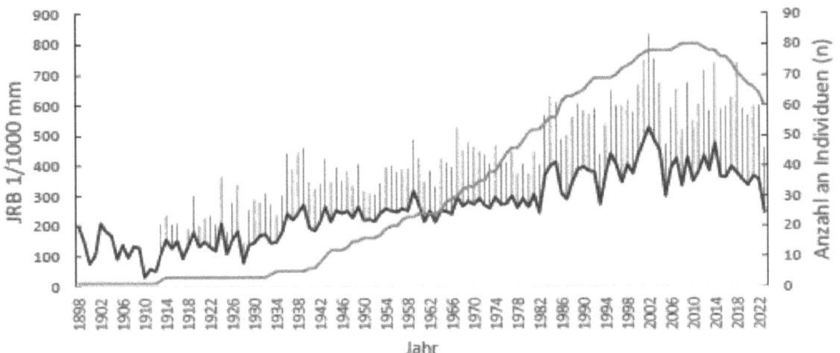

Figure 14: Growth chronology of Juniperus communis and number of samples in the period 1898-2023 (n = 80) (blue line: tree ring width (JRB) with standard deviation (upwards; black); orange line: number of samples)

Figure 15 shows the variation in JRB according to cambial age. For sample numbers > 10, a constant increase can be seen from the cambial age of 5 years, without a pronounced age trend.

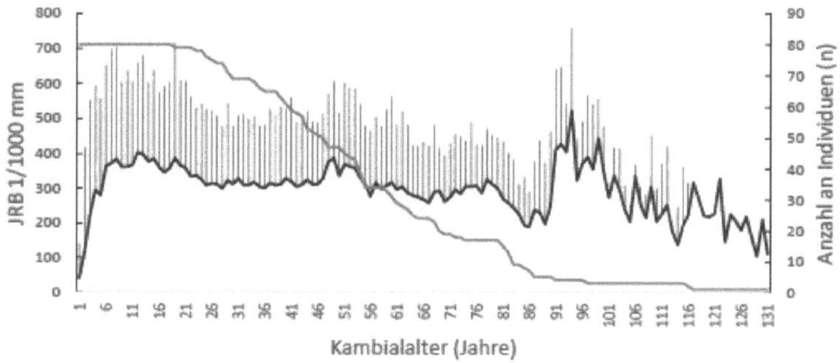

Figure 15: Mean radial growth of Juniperus communis by cambial age blue line: annual ring width (JRB) with standard deviation

Orange line: number of samples

The 80 individuals were then divided into two age groups (Figure 16). For the age group > 50 years, the growth chronology begins in 1901 and ends in 2023. From 1934, the number of samples is over 5 individuals and increases over time to 47 individuals. For the age group < 50 years, the growth chronology begins in 1977 and also ends in 2023. From 1980, the number of samples is equal to 5 individuals and increases to a maximum of 33. For each JRB time series, a mean value was calculated over a total period of 30 years. This period was chosen from 10 to 40 years in order to ensure that the sampled individuals were already 10 years old when the cores were taken (see Figure 17a). The mean value for the chronology of older individuals is 153 pm ± 52 STD, while the chronology of younger individuals is 456 pm ± 80 STD.

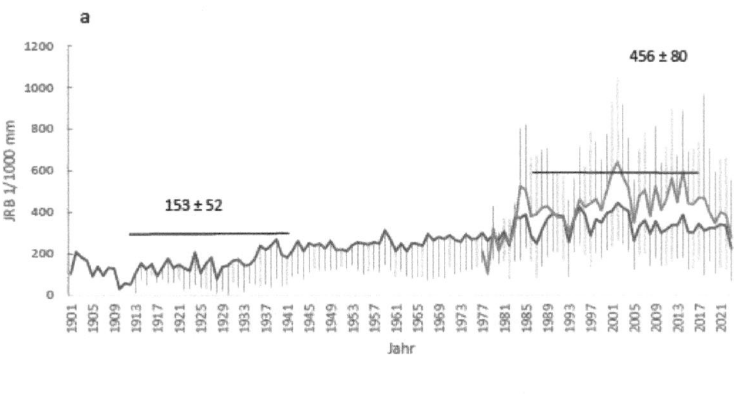

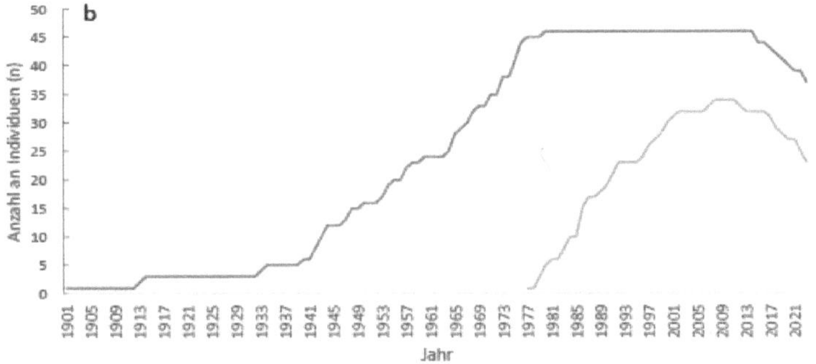

Figure 16: Radial stem growth and number of individuals over the period 1898 - 2023 divided into two age groups. (a) dark blue line: age > 50 years; light blue line: ♦ 50 years; black thin lines: standard deviation {STD); thick black lines: Mean value from JRB of 30 years and the standard deviation
(b) dark orange line: age > 50 years; light orange line: ♦ 50 years

The 80 individuals were also divided into VCs and a mean chronology of radial growth was created for each class (see Figure 17a). Figure 17b shows the number of individuals in all three VCs over the period from 1898 to 2023.

It is noticeable that the values have a pronounced dispersion (high STD), largely show a similar growth trend until 1990 in all UKs, after 1990 a decreasing trend in the growth of UK3 can be observed, i.e. there is an increasing difference in radial growth especially between UK1 and UK3 and the radial growth in all UKs is decreasing in the last decade

25

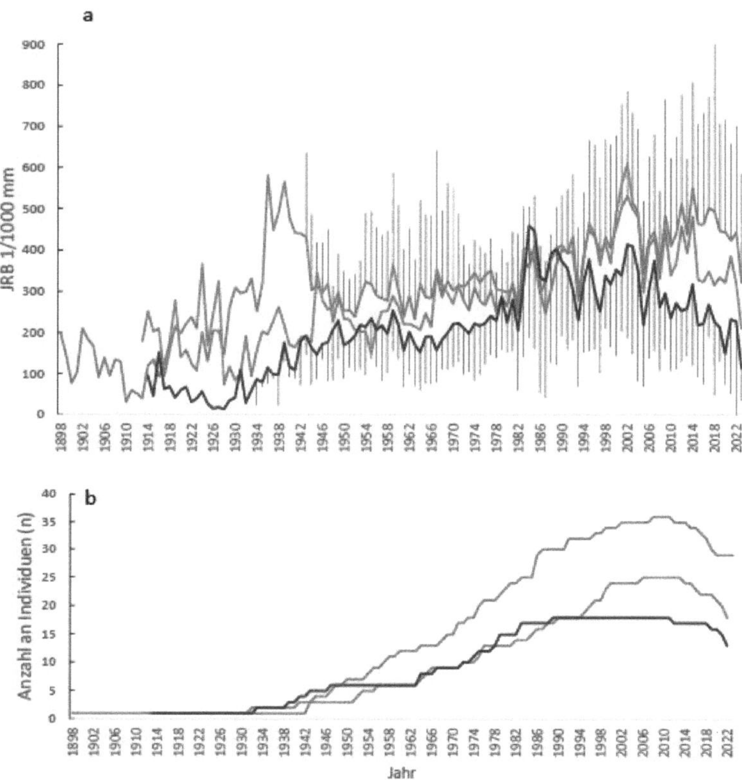

Figure 17: Radial stem growth {a) and number of individuals {b) in the period from 1898 to 2023 divided into three vitality classes {VK). Green: VC 1 {VC < 10 %); orange: VC 2 {VC 10 - 50 %); red: VC 3 {> 50 %); black dOnne lines: standard deviation {STD).

3.4 Base area increase

Figure 18 shows the GFZ of all 80 *Juniperus* communis *individuals* (grey) over the period from 1898 to 2023. The averaged time series shows a slight increase, 0.1 to 0.6 p.m of the GFZ over the period, while the maximum value of an individual is 3.79 p.m and the minimum value is 0.03 p.m. As the number of samples increases, the dispersion of the GFZ also increases.

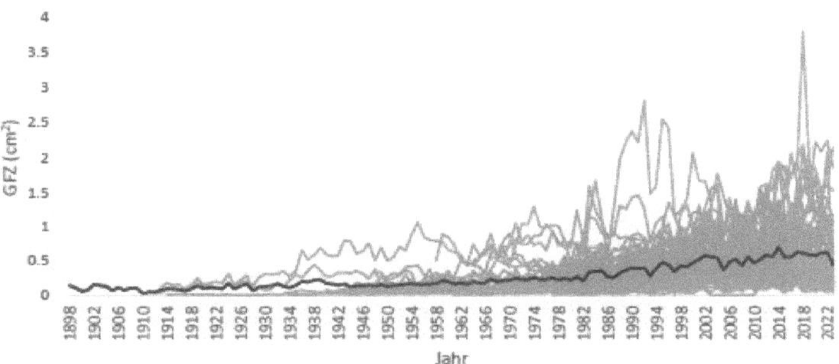

Figure 18: GFZ of all dated Juniperus communis individuals (n=80) and their mean chronology (red line).

Figure 19 shows that the GFZ increases over the period from 1898 to 2023. It starts at 0.14 cm^2 and ends at 0.44 cm^2 . From 1983 onwards, growth rises very sharply and flattens out after 2014. A significant slump in growth can be seen in 1982, 1986, 1987, 1993, 1997, 2005 and 2015, and the chronology rises again in the following year. There is also a slump in growth in 2023.

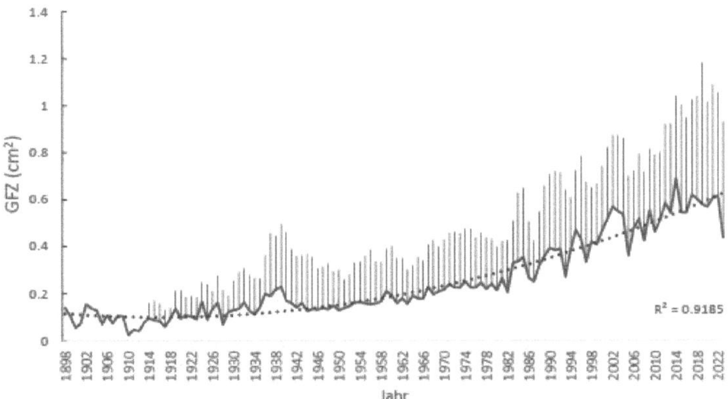

Figure 19: GFZ over the period from 1898 - 2023 (trend line polynomial function; y = 0.00005x^2 - 0.0018x + 0.1158)

The soil depth shows no correlation with the GFZ, the coefficient of determination (R^2) is 0.0002 (see Figure 20). The radial growth of individuals with different crown thinning (= VC 1-3) was also not statistically significantly related to the soil depth surveyed due to the pronounced scattering of the JRB and the GFZ.

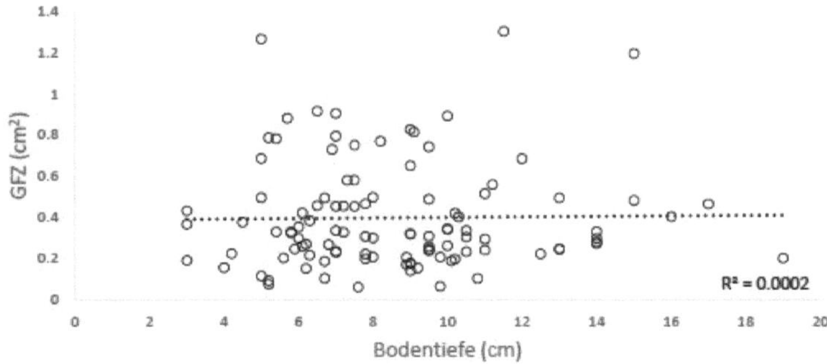

Figure 20: Linear correlation between GFZ and soil depth

The GFZ was also created for the age classes > 50 and < 50 years (Figure 21). The time series of the older individuals show a similar chronology of all 80 individuals. However, the growth slumps are clearer and more pronounced than in the mean chronology of all 80 individuals. The GFZ of individuals < 50 years shows a strong increase until 2014 and decreases in the following years.

The slumps in growth are also clearly recognisable here, but not as pronounced as in the > 50 age group.

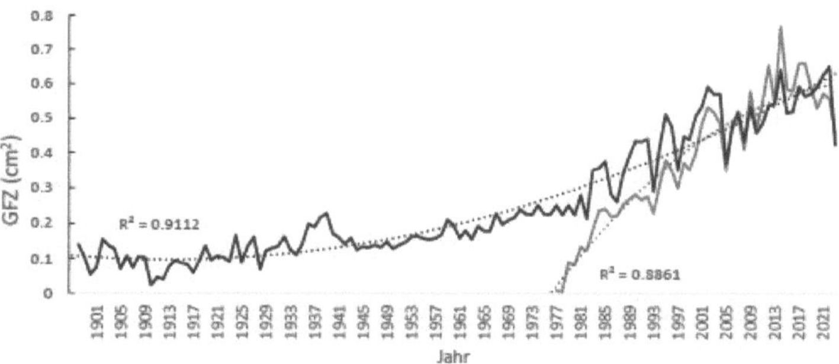

Figure 21: GFZ of both age groups from 1898-2023 (dark blue: > 50 years; y = 0.0005x² - 0.0015x + 0.11 light blue: < 50 years; y = -0.0002x² + 0.0575x - 3.2171; trend lines polynomial function)

The BER of the three CVCs is shown in Figure 22. The CS1 has the highest growth rate and the CS3 the lowest. An increase in all CVs can be observed up to 2003 and, after the growth slump in 2005, there is a further increase in the SPC for CV1 up to 2014; there is a constant trend for CV2 and a decreasing trend for CV3. The extreme growth years are clearly recognisable by the growth slump. The SPC differs statistically significantly (P<0.05) between VC1 and VC2 in the years 2009 and 2010 as well as 20152020, 2022 and 2023. The SPC differs significantly between VC2 and VC3 in the years 1967-1969, 1971, 1973, 1976, 1980, 2001, 2009, 2012- 2014, 2016, 2019-2023.

most pronounced differences (P<0.05) in the GER exist between CSF1 and CSF3, namely in the years 1968, 1974, 1980, 1985, 1991-2006, 2008-2023.

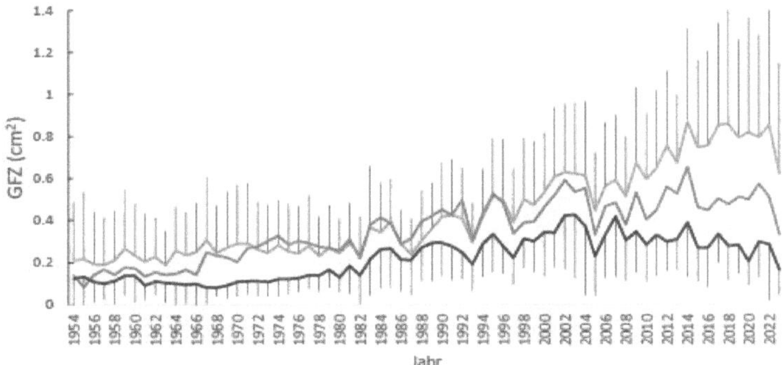

Figure 22: GFZ of Juniperus communis (n: 80) over the period from 1954 to 2023 divided into three vitality classes (VC) (green: VC 1 (KV < 10 %); orange: VC 2 (KV10 - 50 %); red: VC 3 (> 50 %); black and thin lines: standard deviation (STD))

3.5 Climate-growth relationship

Figure 23 shows the correlation coefficient (r) in relation to temperature or precipitation for each month from April of the previous year to September of the current year in a period from 1984 to 2023. In September of the previous year, precipitation has a significant positive influence on stem growth (r= 0.321, p= 0.044).

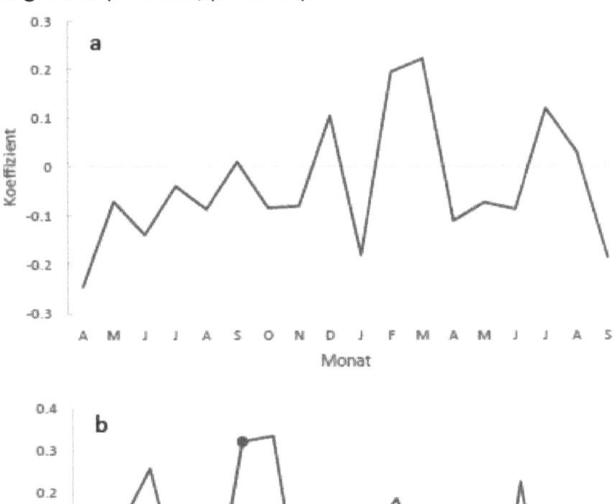

The influence of seasonal climate parameters on radial stem growth was analysed using a linear correlation analysis (Pearson product-moment correlation). The standardised growth chronology was used and the seasonal and monthly mean temperature values and precipitation totals were used as climate parameters. The results of the analysis are shown in Table 5. Stem growth is significantly restricted by precipitation in the previous year during the months of April, May and June (p < 0.05).

Table 4: Person correlation coefficient (r) of the climate-growth relationship between the growth chronology of Juniperus communis and seasonal climate parameters (temperature and precipitation/for the period 1984 to 2023 (Fr = spring; AMJ = April, May, June; So = summer; Her = autumn; Wi = winter Red: significant (p <0.05))

Correlation coefficient	1984-	2023						
	Previous year					Latest news	Year	
Temperature	Fri	AMJ	So	Her	Wi	Fri	AMJ	So
r	-0.134	-0.219	-0.13	-0.088	0.125	-0.02	-0.129	0.013
P	0.436	0.174	0.423	0.591	0.442	0.901	0.428	0.935
Precipitation								
r	0.168	0.333	0.029	0.115	0.06	-0.024	-0.791	-0.004
P	0.301	0.036	0.857	0.478	0.715	0.134	0.628	0.979

Figure 24 shows a moving correlation curve between precipitation from April to June and growth over the period from 1951 to 2023. Figure 24a shows the correlation between growth and precipitation in the current year, while 24b shows that of the previous year.

The sliding correlation between growth and the April to June precipitation of the current year does not correlate significantly. However, the influence of precipitation from April to June of the previous year shows an opposite trend (Figure 24b) and shows a statistically significant influence on radial growth in the last periods.

Figure 24: Sliding correlation between precipitation April-June and radial growth of Juniperus communis (period 1951-202); dashed line: up to 5 n solid line: n above 5 filled circle: significant correlation (P<0.05) (a) Correlation between growth and precipitation April-June of the current year (b) Correlation between growth and precipitation April-June of the previous year

Figure 25 also shows correlation curves. One for the spring temperature (March - May) of the current year and one for the summer temperature (June - August) of the current year. As can be seen in Figure 25a, there is no correlation between growth and the spring temperature of the previous year. Increasing trend in correlation to summer temperature, although not yet significant.

Figure 25: Sliding correlation curves between radial growth and temperature of the current year of Juniperus communis from 1951 – 2023; dashed line: up to 5 n solid line: above 5 n
(a) Correlation between growth and mean temperature March to May
(b) Correlation between growth and mean temperature June to August

3.6 Extreme slumps in growth

The identification of extreme years was based on the GFZ for the period from 1934 to 2023 (see Figure 26). Extreme growth slumps and recoveries are shown in Figure 26. The sharpest slumps in growth occurred in 1993, 1997, 2005 and 2023 (2005: -1.16 STD).

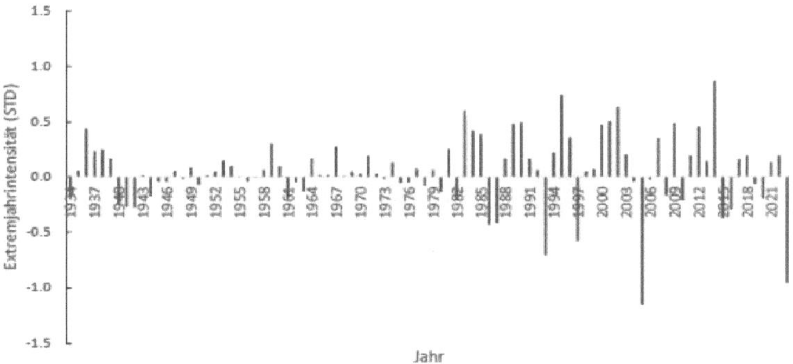

Figure 26: Extreme year determination from the period 1934 to 2023 based on the GFZ of Juniperus communis (n = 80) negadve/posidve values mean a growth deviation ranging from -1.5 to +1.5 STD

The extreme negative growth years were determined separately for all three UKs (see Figure 27). There are some differences between the three UKs. If the threshold value for assignment to a negative extreme growth year is set at 0.5 STD, then the extreme growth years 1987, 1993, 1997, 2005 and 2023 occur in VC1, the years 1986, 1993, 1997, 2005 and 2023 in VC2 and the years 1993, 1997, 2005, 2010, 2015, 2016, 2020 and 2023 in VC3. VK3 has the highest number of extreme negative growth years. The slump in growth is also more pronounced in UK 3 than in the other two, although it is lowest in UK 1. The extreme year 2005 shows the sharpest drop in growth for all classes, followed by the year 2023.

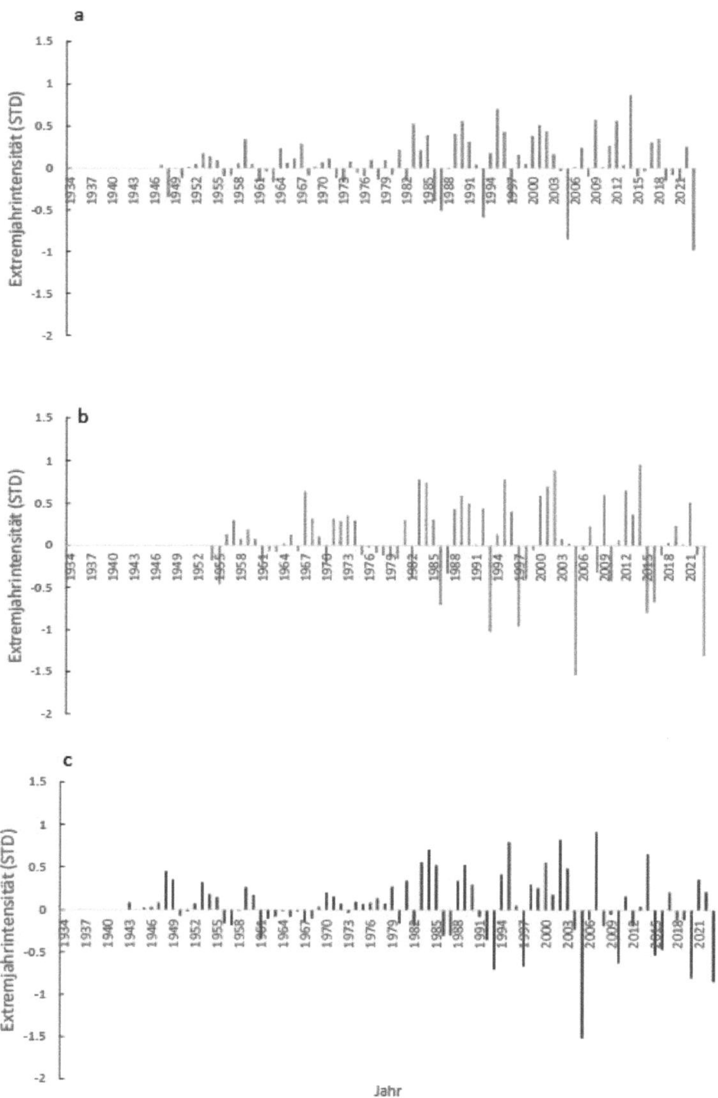

Figure 27: Extreme year determination from the period 1934 to 2023 based on the GFZ of Juniperus communis in Dependence on the UK
negadve/posidve values mean a growth deviation in standard deviation units ranging from -2 to +1.5 (a)VKI (b)VK2(c)VK3

Table 6 shows the temperature and precipitation values from April to June for individual extreme growth years (1986, 1993, 2005, 2015 and 2023). The period from 1994 to 1997 is also shown in order to analyse the recovery. The precipitation amounts are low in the previous year of an extreme growth year.

Table 5: Precipitation and temperature values (April-June, spring, summer) in extreme growing seasons
Light blue: precipitation below the long-term average; dark blue: precipitation above the long-term average; light red:
temperature above the long-term average

Growing years	Apr-JuneT	Apr-JuneNS	FrNS	SoNS
MW	11.8	200	129	321
1985	11.2	200.0	117.0	182.0
1986	12.1	209.0	95.0	439.0
1992	12.1	159.6	116.8	286.8
1993	13.2	185.0	189.5	248.4
1994	11.9	212.9	117.6	375.2
1995	11.0	209.3	131.2	266.5
1996	12.5	182.1	138.3	363.3
1997	11.0	223.2	141.0	343.8
2004	11.6	94.3	121.8	297.2
2005	13.2	166.7	94.4	195.1
2014	13.1	231.9	153.8	258.8
20151	12.8	273	156.2	363.1
2022	14.2	251.7	107.7	420.5
2023	13.5	249.9	119.9	345.5

3.7 Stress indices in extreme growing years

The stress indices - resistance, recovery and resilience - were then presented for the three UKs for the years 1986, 1993 and 2005. These extreme years were selected because they cover a longer period and these years occur with varying intensity in all UKs.

Figure 28 summarises the data, with outliers removed. Values that were much too high or too low than the normal range of the respective CV were removed. In all three CVs, there were a few individuals with higher/lower resistance, recovery and resilience than all others. The appendix (see Figure A5) contains graphs in which all individuals are included.

In 1986, it can be seen that the resistance of VK 1 is the highest with a value of 0.84, while the other two classes are fairly equal, VK 2 0.76 and VK 3 0.79. In 1993, all values are also below 1 and are quite close to each other, VK 1 0.7, VK 2 0.74 and VK 3 0.69, and in 2005 the values are also below 1, with VK 1 having the highest value at 0.72, VK 2 0.61 and VK 3 the lowest at 0.57.

The recovery is above 1 for all three UKs, which means that they show a good recovery. In 1986, the recovery of UK 3 is the highest, 1.25, and UK 1 and UK 2 are above 1 (1.046 and 1.072). In 1993, all values are higher than in 1986. UK 1 is the highest with a value of 1.59, UK 2 has a value of 1.43 and UK 3 has a value of 1.38. In 2005, all UKs also show a good recovery, with UK 3 having the highest value of 1.44. UK 2 shows a slightly lower recovery with a value of 1.43 and UK 1 is the lowest with a value of 1.24.

In 1986, the resilience of UK 3 is above 1 with a value of 1.17, while UK 1 is significantly lower at 0.89 and UK 2 at 0.9. In 1993, all values of the UKs are above 1, UK 1 1.14, UK 2 1.149 and UK 3 1.16. In 2005, all values of the UKs are again below 1, UK 1 0.91, UK 2 0.84 and UK 3 0.88.

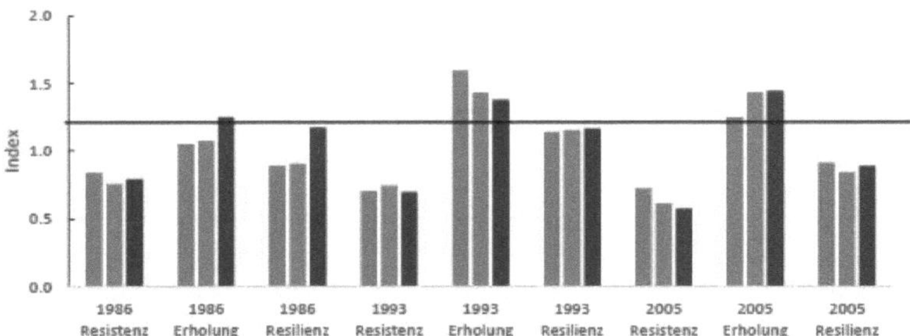

Figure 28: Stress indices (resistance, recovery and resilience for the years 1986, 1993 and 2005) based on the Basal area increment of Juniperus communis as a function of vitality class (VK) (green: VK 1; orange: VK 2; red: VK 3; black line: limit value 1)

3.8 Stem growth of *Juniperus communis* compared to *Pinus sylvestris*

The *Pinus sylvestris* growth chronology (n= 100) of Drexler (2020) is quite similar to the *Pinus sylvestris* chronology of this study (n= 10). Both chronologies were compared as the time series in this study was extended to 2023 (see Figure 29). Table 7 shows the highly significant overlap in growth between these two time series. The synchronisation in the growth fluctuations of the *Pinus sylvestris* chronologies can also be clearly seen in Figure 30.

The SPC of *Pinus sylvestris* differs markedly from that of *Juniperus communis* (Figure 30). In the period 1901-2023, *Juniperus communis* has a mean GFC of 0.0 to 0.7 cm^2 , while *Pinus sylvestris* has a GFC of 1.3 to 2.8 cm in the same period.^{2}
shows.

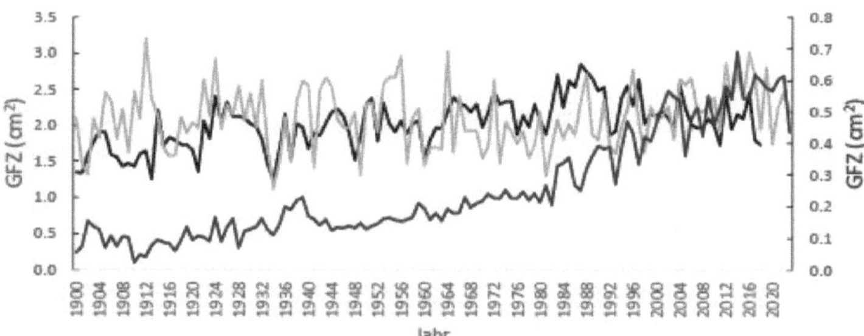

Figure 29: Comparison of the GFZ of Pinus sylvestris and Juniperus communis for the period 1900-202.
(Black line: GFZ Pinus sylvestris, Drexler (2020); turquoise line: GFZ Pinus sylvestris, this study; blue line: GFZ Juniperus communis, this study)

The *Pinus sylvestris* data *of* this study, designated as 241Pi_m, were checked for synchrony in growth chronology with those of Drexler (2020), designated as K241_MKn78 (100 individuals) (see Table 7). The calculated synchronisation (GLK) shows that the growth chronology of *Pinus sylvestris* in this study shows a synchrony of 71% (P<0.001) in the overlap area with the chronology of Drexler (2020). TheTVBP value of 6.1 indicates a high degree of synchronisation, as a value above 4.5 indicates a strong overlap.

The growth chronology of *Juniperus communis* shows no significance in comparison with that of *Pinus sylvestris* (this study) in the overlap area (n= 126) (Table 5). However, the GLK indicates that the chronologies are not very similar, with a similarity of 52 %. The TVBP value is very low at 0.9, which indicates that the two time series are not synchronised. This can also be seen in Figure 29.

The growth chronology of Drexler's study (2020) shows no significance compared to that of *Juniperus communis* in the overlap area (n = 121) (Table 5). The two chronologies have a GLK of 54%, which indicates that they are not very similar. The TVBP value of 1.4 is low and shows that the chronologies are not very synchronised.

Table 6: Statistical parameters of the synchronisation between the Pinus sylvestris chronology and the Juniperus communis chronology of this study and the Pinus sylvestris chronology of Drexler (2020) K241_MKn78 of Drexler (2020) 241Pi_m: Time series of this study (n = 10)   K241_MKn78: Time series Drexler (2020); n = 100)) Jucon80: Juniperus communis (n = 80); OVL = overlap length, GLK = synchronisation, TVBP = t-value Baillie and Pilcher, DateL = start date, DateR = end date

Sample	Ref.	OVL	GLK		TVBP	DateL	DateR
241Pi_m	K241_MKn78	141		71***	6.1	1878	2023
Jucon80	241Pi_m	126	52		0.9	1898	2023
Jucon80	K241_MKn78	121	54		1.4	1898	2023

4 Discussion

4.1 Radial and ground surface growth

As part of this master's thesis, the radial growth of tree-form *Juniperus* communis *individuals* in the vicinity of the FAIR research station in Mieming was investigated by means of a dendroocological study.

In the course of this work, it was found that there is no significant correlation between stem diameter and stem height, stem height and soil depth, stem diameter and soil depth, age and stem height, and age and stem diameter. The large variation in the stem parameters (height, diameter, JRB) indicates pronounced small-scale differences in shading by the pine stand as well as the nutrient and/or water supply. The large variation in the soil depths surveyed supports this assumption. *Juniperus communis* normally grows up to 5 metres high, but can also reach up to 17 metres (Gilbert, 1980). Possible causes for such different sizes may be shade, grazing and, in the highland mountains, exposure (Gilbert, 1980). This explains why stem height and stem diameter are not correlated, as are stem height and tree age. The radial growth of individuals with different crown thinning (= VK 1-3) was also not statistically significantly related to the soil depth surveyed due to the pronounced dispersion of the JRB.

Soil depth was measured in order to use it as an indirect measure of nutrient and water availability. However, it must be noted that it is unknown how far the root system of each individual reaches and therefore where it can obtain nutrients. *Juniperus communis* often grows on soil with a low nitrogen content and has an indicator value of 3 (Hill et al., 1999). The plant is therefore very well adapted to low-nutrient conditions and is limited more by the availability of light than by nutrients (Grubb et al., 1996).

The JRB chronology shows a slight increase in the JRB over the years. In 1920 it had an average width of 146 pm and in 2022 one of 354 pm. However, it must be taken into account that there is an age trend and the number of individuals varies over the period. If the JRB are arranged according to cambial age, a slight decrease in the JRB can be seen with increasing age of the trees. At the age of 7 years the average width is 372 pm, whereas at the age of 130 years it is only 212 pm. Initially, trees form wider growth rings, which then become increasingly narrower (Kindermann & Neumann, 2011).

A JRB chronology was also divided into the age classes < 50 years and > 50 years. At first glance, it can be seen that the younger Juniperus *individuals* grow significantly better than the older ones. At an age of 10-40 years, the individuals >50 years have a JRB of 153 pm, while the younger ones have a mean value of 456 pm. From 1970 onwards, however, a slight increase in the JRB can also be recognised in the older Juniperus *individuals* and a stronger increase from 1980 onwards. This is most likely due to the fact that the area was used for grazing cattle before 1970. Grazing was then discontinued and the stand was able to recover and expand, as can be seen in Figure 16. In the *Juniperus* communis *stand,* the cattle use resulted in constant browsing and trampling, which caused many specimens to die off. Without this disturbance, *Juniperus communis* was able to develop freely in the following years. *Juniperus communis* that grew from this point onwards were able to develop much better. Cattle and sheep gnaw on the bark of *Juniperus communis* and thus significantly influence its shape and size. In extreme cases, the animals can open, fragment or completely destroy a dense *Juniperus communis* stand (Huntley & Birks, 1979a; Gilbert, 1980; Borders Forest Trust, 1997). *Juniperus communis* has a high content of monoterpenoids, which can cause digestive and kidney problems, miscarriages and even death in goats and cattle (Gardner et al., 1998; Philip, 2003). Nevertheless, many large mammals, including deer, elk and horses, are registered as eating it, usually when other food is scarce or of poor quality (Thomas & Polwart, 2003).

The GFZ was then displayed graphically. This clearly shows that the GFZ rises sharply from 1982 and reaches a plateau in 2014. As soil depth is an indirect measure of nutrient availability, it can be seen that nutrient availability is not related to the GFZ. However, it must be noted here that it is not known how far the rooting of the individual individuals extends.

As with the JRB, the GFZ also shows a clear increase, especially when comparing the younger and older *Juniperus* individuals. This sudden increase in the GFZ can be explained by the cessation of grazing and utilisation. The plateau is reached when the plants begin to compete for light, nutrients and water. Figures 19 and 21 show that the density of the pine stand increased significantly between 1970/74 and 2023. This can be attributed to changes in land use (reduction or abandonment of grazing and/or possibly also timber utilisation) during this period. The above- and below-ground root system of *Juniperus communis* was severely affected by browsing and trampling damage. As a result of the change in land use, the stands recovered and the GFZ rose sharply from 1983 onwards.

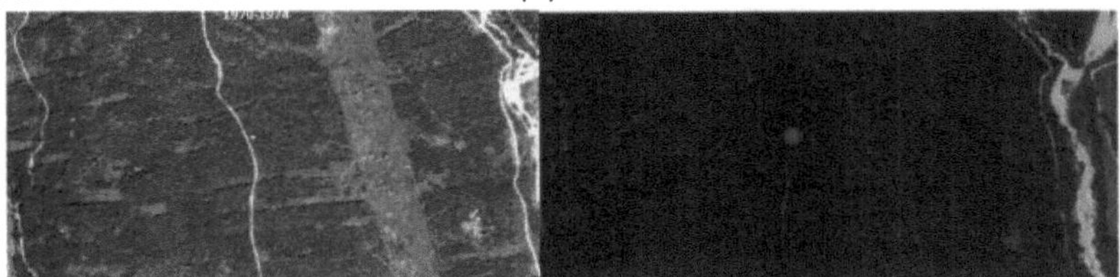

Figure 30: Aerial photographs of the study area
black and white aerial photograph from 1970-1974 (TIRIS, Province of Tyrol,
https://www.tirol.gv.at/sicherheit/geoinformation/geodaten- tiris/orthofotos/)
Colour aerial photo from 2023 (https://earth.google.com)
red dot: FAIR forest station in Mieming

A similar pattern can be observed in the subdivision into the three VCs: In recent years, CC 1 has had a wider JRB than CCs 2 and 3, with CC 3 having the narrowest rings. This becomes even clearer when looking at the total area growth (TAG). From 2008 onwards, CC 1 shows a steady increase in the BER, while CC 2 remains at a constant level and CC 3 shows a decline, which was to be expected

The second hypothesis, that the GFZ decreases in the course of climate change, applies to individuals with already reduced vitality (i.e. pronounced crown thinning VK3). From the hot summer of 2003 onwards, there was a negative trend in the GFZ in these individuals and the GFZ deviated significantly ($P<0.05$) from the GFZ of individuals with high vitality (VK1) in the following years.

4.2 Climate-growth relationship

The April-June precipitation remained largely constant in the period 1951-2023, while the April-June temperature increased by around 2 °C in this period (Figure 8). The correlation graphs and the stability graph show that the precipitation of the current year has no significant influence on radial stem growth. However, the precipitation of the previous year became significant in the last six years. The spring and summer temperatures of the current year also have no significant influence on radial stem growth. Thus, the first hypothesis could only be partially confirmed. Radial growth is not limited by temperature and precipitation together, but mainly by the precipitation (April to June) of the previous year.

The genus *Juniperus* has a high hydraulic safety (Unterholzner et al. 2020; Beikircher and Mayr 2008; Mayr et al. 2006; Willson and Jackson 2006). *Juniperus communis* is known to have a slow growth rate, high longevity and a high ability to cope well with low temperatures, water stress, strong winds and food scarcity (Unterholzner et al. 2020; Adams 2014; Thomas et al. 2007). *Juniperus communis* has a wide distribution range, growing from sea level in coastal areas to the Krummholz belt above the tree line, suggesting that it can easily adapt to a wide range of conditions.

The significant influence of precipitation in recent decades on the radial growth of the plant indicates that water is becoming a strong limiting factor due to climate change.

4.3 Extreme growth slumps and stress indices

When identifying the extreme years, it was clear to see that the most extreme growth slumps were recorded in 1993, 1997, 2005 and 2023. This analysis was also carried out for the three UKs and showed that Vit 1 has only 3 extreme growth years: 1987, 2005 and 2023, Vit 2 has 7 extreme growth years: 1986, 1993, 1997, 2005, 2015, 2016 and 2023 and Vit 3 has 8 extreme growth years: 1993, 1997, 2005, 2010, 2015, 2016, 2019 and 2023.

The problem, however, is that only monthly precipitation amounts were available. This means that heavy precipitation within a few days can represent the entire amount of precipitation for a month. There is only a shallow soil depth at the site, which rests on the rock, causing the water to run off and seep away quickly. This leads to a low water storage capacity, and thus the actual water availability during the growing season deviates from the monthly precipitation amounts. The available soil water is decisive for plant growth, not the amount of precipitation (Oberhuber et al. 1998). It therefore remains unclear how much water was actually available to the plants.

Nevertheless, a good interpretation is possible, as both monthly precipitation amounts and temperature totals are available and the extreme growth years were determined using the GFZ.

The higher temperature, or during a drought or dry period, does not directly affect growth, but does affect evaporation. Water is released from the ecosystem into the surrounding air, which is known as evapotranspiration. This means that some of the water is lost and is no longer available to the plants. The drier the air, the more water is released into the air, especially during periods of drought, as the air is dry and warm (Beer et al. 2009; Dai et al. 2018).

Evapotranspiration is expected to increase even further in the future due to climate change (Beer et al. 2009; Dai et al. 2018). Periods of drought are mainly recorded in spring and summer, i.e. during the above-ground growth phase of *Juniperus communis* and *Pinus sylvestris.* Some studies show that an improved water utilisation efficiency could occur in plants with increased CO2 levels, as they could reduce evaporative demand by reducing stomatal aperture width and thus curb desiccation (Cheng et al. 2017).

The stress indices (resistance, recovery and resilience) for the individual UKs were determined for the years 1986, 1993 and 2005.

The low resistance values of *Juniperus communis* in the selected years are probably due to the direct limitation of cambium activity as a result of limited water availability (Korner, 2015). Photosynthesis may also be reduced, thereby reducing growth (Herrero and Zamora, 2014; Fang and Zhang, 2020).

However, *Juniperus communis* shows a rapid recovery of radial growth after a drought period, sobaid the water resources are available again. Growth can then be resumed. This shows that *Juniperus communis* has a flexible utilisation of water resources and an effective restoration of the water balance (Wang et al., 2019; Fang and Zhang, 2020; Shao et al., 2020; Piraino et al., 2024).

Low resilience indicates that plants are able to recover quickly after drought events, but not fully. Resilience is not only a measure for survival after a disturbance, but also for the recovery of the original growth pattern. Thus, *Juniperus communis* probably needs a little longer to return to its original growth level (Wang et al., 2019; Piraino et al., 2024). Thus, *Juniperus communis* tolerates droughts very well, although the strongest droughts can also affect it (Thomas et al., 2007). When plants are exposed to CO2 levels of 360 to 700 ppm, this has an impact on water use efficiency (Togneffi et al., 2002). They show a lower stomatal conductance of the leaves, which means that they lose less water through transpiration. However, this also reduces photosynthesis, leaf cooling and nutrient uptake. Nevertheless, tugor pressure can be maintained osmotically, even during drought. With a high CO2 content in the atmosphere, the plant can carry out more photosynthesis, which leads to greater carbohydrate synthesis and an accumulation of osmotically active dissolved substances in the leaves.

leaves. This in turn leads to a reduction in leaf water potential (Togneffi et al., 2002). As a result, *Juniperus communis* was able to develop greater drought tolerance when CO2 levels increased (Thomas et al., 2007). Due to the increased leaf water potential, it can also absorb water better from the soil (Thomas et al. 2007). *Juniperus communis* is very drought-tolerant, as it has only a low proportion of xylem embolisms even under drought conditions (Togneffi et al., 2001).

4.4 Comparison of the growth chronology of *Juniperus communis* and *Pinus sylvestris*

The year-to-year variations in radial growth between the dominant woody species in the tree and shrub layer, i.e. *Pinus sylvestris* and *Juniperus communis,* show no significant correlation ($p > 0.05$). It can therefore be concluded that different climatic factors limit the radial growth of these species. *Pinus sylvestris* has also exhibited an almost constant GFZ of 2.26 ± 0.38 cm^2 since the beginning of the 20th century. In contrast, *Juniperus communis* shows at the beginning of the 20th century. GFZ values of only 0.1 cm^2 , which rise to a maximum of 0.7 cm^2 in the last decade. These marked differences in radial growth indicate the competitive strength of *Pinus sylvestris.*

The climate-growth correlations of *Juniperus communis* indicate a low sensitivity of radial growth to climatic factors, but the massive limitation of growth due to land use (presumably grazing until 1970) does not allow a definitive statement regarding sensitivity to drought stress. In addition, sliding climate-growth correlations show an increasing limitation of the radial growth of *Juniperus communis* to precipitation in spring.

Based on these surveys, hypothesis 3, *Juniperus communis* has higher drought resistance than *Pinus sylvestris*, cannot be confirmed.

5 Conclusions

In conclusion, it can be stated that the reduction in utilisation intensity after around 1970 not only favoured stand development, but also the radial and shoot growth of tree-form *Juniperus communis* individuals due to the decrease in trampling and browsing damage. In recent decades (i.e. the 30-year periods from 1988-2017), however, radial growth has been significantly limited by precipitation intensity in April to June of the previous year. Furthermore, climate warming combined with limited nutrient and water availability led to an increase in root competition with *Pinus sylvestris*, whereby *Juniperus* communis *individuals* with low vitality show a constant decrease in the GFZ after 2005.

Based on the comparison of the GFZ (long-term trends and absolute values) between *Pinus sylvestris* and the individual vitality classes of *Juniperus communis* as well as the sliding climate-growth correlations for *Juniperus communis*, it can be assumed that *Pinus sylvestris* has a higher drought resistance than *Juniperus communis* in the study area. However, in order to clarify this in detail, ocophysiological surveys (e.g. leaf water potential, photosynthesis, transpiration, intra-annual growth) are required during and after pronounced dry periods for these species.

6 Bibliography

Adams, R. P. (2014). *Junipers of the world: the genus Juniperus.* Trafford Publishing, Bloomington, IN, USA.

Anchukaitis, K.J. (2017): Tree rings reveal climate change past, present, and future. *Proceedings of the American Philosophical Society* 161(3): 244-263.

Auer I., Bohm R., Jurkovic A., Lipa W., Orlik A., Potzmann R., Schoner W., Ungersbock M., Matulla C., Briffa K., Jones P.D., Efthymiadis D., Brunetti M., Nanni T., Maugeri M., Mercalli L., Mestre O., Moisselin J.M., Begert M., Muller-Westermeier G., Kveton V., Bochnicek O., Stastny P., Lapin M., Szalai S., Szentimrey T., Cegnar T., Dolinar M., Gajic-Capka M., Zaninovic K., Majstorovic Z., Nieplova E. (2007): HISTALP - historical instrumental climatological surface time series of the Greater Alpine Region 1760-2003. *International Journal of Climatology* 27, 17-46, doi:10.1002/joc,1377.

Baumgarten, B. (2014): CARLOS DE GIMBERNAT'S FIRST GEOLOGICAL MAP OF TYROL (1808) CARLOS DE GIMBERNAT AND THE FIRST GEOLOGICAL MAP OF TYROL (1808). Geo.Alp.

Bartels, H. (1993): *Geholzkunde. Introduction to dendrology.* Ulmer, Stuttgart 1993, ISBN 38252-1720-5.

Beer, C., Ciais, P., Reichstein, M., Baldocchi, D., Law, B. E., Papale, D., Soussana J. F., Ammann C., Buchmann N., Frank D., Gianelle D., Janssens I.A., Knohl A., Kostner B., Moors E., Roupsard O., Verbeeck H., Vesala T., Wiliams C.A. & Wohlfahrt, G. (2009). Temporal and among-site variability of inherent water use efficiency at the ecosystem level. *Global biogeochemical cycles,23*(2). doi: 10.1029/2008GB003233.

Beikircher B, Mayr S (2008) The hydraulic architecture *of Juniperus communis* L. ssp. communis: shrubs and trees compared. *Plant Cell Environ* 31:1545-1556.

Biondi F. (1997): Evolutionary and moving response functions in dendroclimatology. Dendrochronologia 15:139-150.

Biondi F. and Waikul K. (2004): DENDROCLIM 2002: a C++ program for statistical calibration for climate signals in tree-ring chronologies. Comput Geosci 30: 303-311.

Biasing TJ, Solomon AM, Duvick DN (1984) Response Functions Revisited. Tree-Ring Bulletin, 44, 1-15.

Bohlmann D. (2013): *Wood biology. Why Баўme do not grow into the sky.* 2nd edition. Quelle & Meyer, Wiebelsheim 2013, ISBN 978-3-494-01547-7.

Borders Forest Trust (1997) Common juniper (*Juniperus communis* L.): a review of its biology and status in the Scoffish Borders. *Borders* Forest Trust Occasional Paper No. 1, Ancrum, Roxburghshire.

Bohm R. (ed.) (2006): *ALP-IMP. Multi-centennial climate variability in the Alps based on instrumental data, model simulations and proxy data.* Vienna: Central Institute for Meteorology and Geodynamics, project report, 102 pages.

Bohm R. (2010): *Heifie Luft - nach Kopenhagen. Climate change. Facts - Fearful business.* 2nd ed. Vienna, Klosterneuburg: Edition Va Bene, 280 pages, ISBN 978-3851672435.

Buntgen U., Esper J., Frank D.C., Nicolussi K., Schmidhalter M. (2005): A 1052-year tree-ring proxy for Alpine summer temperatures. Climate Dynamics 25, 141-153, doi:10.1007/s00382-005-0028-1.

Buntgen U., Frank D.C., Nievergelt D., Esper J. (2006): Summer temperature variations in the

European Alps, AD 755-2004, Journal of Climate 19, 5606-5623, doi:10.1175/JCLI3917.1.

Cheng, L., Zhang, L., Wang, YP. *et al.* Recent increases in terrestrial carbon uptake at little cost to the water cycle. *Nat Commun* 8, 110 (2017). doi:10.1038/s41467-017-00114-5.

Cook, E. R., Seager, R., Kushnir, Y., Briffa, K. R., Buntgen, U., Frank, D.,... & Zang, C. (2015): Old World megadroughts and pluvials during the Common Era. *Science advances*,1(10), el500561, doi:10.1126/sciadv,1500561.

Dai, A., Zhao, T., & Chen, J. (2018). Climate change and drought: a precipitation and evaporation perspective. *Current Climate Change Reports, 4,* 301-312.

Drexler L. (2020): Resilience of a pine stand on the Mieminger Plateau (Tyrol) to drought stress. Master thesis, Leopold Franz University of Innsbruck, 76 pages.

Ellenberg H, Leuschner C (2010) Vegetation of Central Europe with the Alps. From an ecological, dynamic and historical perspective. UTB, Stuttgart, 1095 pp.

Fang, O., Qiu, H., & Zhang, Q. B. (2020). Species-specific drought resilience in juniper and fir forests in the central Himalayas. *Ecological Indicators, 117,*106615.

Fluxnet Austria. (2024). University of Innsbruck. Retrieved from [Link to the website of the University of Innsbruck].https://www.uibk.ac.at/fakultaeten/biologie/alpine-forschungsstaetten/obermieming.html.de.

Fritts, H.C. (1976). Tree Rings and Climate Academic Press. New York.

Fritsch, M. (2012). Vegetation-ocological studies on the management and restitution of wet grassland. Dissertation, TU Darmstadt, D.

Gardner, D.R.,Panter, K.E.,James, L.F.& Stegelmeier, B.L.(1998) Abortifacient effects of lodgepole pine *(Pinus contorta)* and common juniper *(Juniperus communis)* on cattle. *Veterinary and Human Toxicology,* 40, 260-263.

Gilbert, O.L. (1980) Juniper in Upper Teesdale. *Journal of Ecology,* 68, 1013-1024.

Grubb, P.J.,Lee, W.G.,Kollmann, J.& Wilson, J.B. (1996) Interaction of irradiance and soil nutrient supply on growth of seedlings of ten European tail-shrub species and *Fagus sylvatica.JournalofEcology,* 84, 827-840.

Gruber A, Strobl S, Veit B, Oberhuber W (2010) Impact of drought on the temporal dynamics of wood formation in *Pinus sylvestris.* Tree Physiology, 1-12.

Herrero, A., & Zamora, R. (2014). Plant responses to extreme climatic events: a field test of resilience capacity at the southern range edge. *Pios one, 9*(1), e87842.

Hill, M.O., Mountford, J.O., Roy, D.B. & Bunce, R.G.H. (1999) *Ellenberg's Indicator Values for British Plants.* ECOFACT, Vol. 2, Technical Annex. Institute of Terrestrial Ecology, Huntingdon, UK.

Holmes, R.L. (1983). Computer-assisted quality control in tree-ring dating and measurement. Tree-Ring Bulletin, 43, 69-78.

Huntley, B. & Birks, H.J.B. (1979a) The past and present vegetation of the Morrone Birkwoods National Nature Reserve, Scotland: I. A primary phytosociological survey.*Journal of Ecology,* 67, 417-446.

IPCC (2021): Summary for Policymakers. In: Climate Change 2021: The Physical Science Basis. Contribution of Working Group I to the Sixth Assessment Report of the Intergovernmental Panel on Climate Change [MassonDelmotte, V., P. Zhai, A. Pirani, S.L. Connors, C. Péan, S. Berger, N. Caud, Y. Chen, L. Goldfarb, M.I. Gomis, M. Huang, K. Leitzell, E. Lonnoy, J.B.R.

Matthews, T.K. Maycock, T. Waterfield, O. Yelekgi, R. Yu, and B. Zhou (eds.)]. Cambridge University Press, Cambridge, United Kingdom and New York, NY, USA, pp. 3-32, doi:10.1017/9781009157896.001.

Jagel, A. (2012). Secretly and often unnoticed: The conifers are bleeding. *Jahrb. Bochumer Bot.*

Ver, 3, 227-232.

Karanitsch-Ackerl, S., Mayer, K., Gauster, T., Laaha, G., Holawe, F., Wimmer, R., & Grabner, M. (2019): A 400-year reconstruction of spring-summer precipitation and summer low flow from regional tree-ring chronologies in North-Eastern Austria.*Journal of Hydrology,*577, 123986, doi:10.1016/j.jhydrol.2019.123986.

Kindermann, G., & Neumann, M.(2011) Radial increment changes over time based on core analyses. *Section Yield Science Annual Conference 2011 Cottbus,* 120.

Korner, C. (2015) Paradigm shift in plant growth control. *Current Opinion in Plant Biology* 25, 107-114.

Mandi, G. W. (1996). On the geology of the Odenhof Window (Northern Limestone Alps, Austria). *Yearbook of the Geological Survey of Austria, 139*(4), 473-495.

Mayr S, Hacke U, Schmid P, Schwienbacher F, Gruber A (2006) Frost drought in conifers at the alpine timberline: xylem dysfunction and adaptations. *Ecology* 87:3175-3185.

Muller-Using, S. (2005). The Scots pine: *Pinus sylvestris*. Stuttgart: Ulmer Verlag.

Pichler P, Oberhuber W (2007) Radial growth response of coniferous forest trees in an inner Alpine environment to heat-wave in 2003, Forest Ecology and Management, 242, 688-699.

Piraino, S., Hadad, M. A., Ribas-Fernandez, Y. A., & Roig, F. A. (2024). Sex-dependent resilience to extreme drought events: implications for climate change adaptation of a South American endangered tree species. *EcologicalProcesses, 13*(1), 24.

Unterholzner, L., Carrer, M., Bar, A., Beikircher, B., Damon, B., Losso, A., ... & Mayr, S. (2020). *Juniperus communis* populations exhibit low variability in hydraulic safety and efficiency. *Tree Physiology, 40*(12), 1668-1679, doi: 10.1093/treephys/tpaal03.

Roloff, A., Bartels, A. (2006): *Flora der Geholze: Bestimmung, Eigenschaften und Verwendung.* Eugen Ulmer, Stuttgart 2006, ISBN 3-8001-4832-3.

Roloff, A. Weisgerber, H., Ulla M. Lang, Bernd Stimm (eds.) (2007): *Encyclopaedia of Wood Waxes. Handbook and atlas of dendrology.* Wiley-VCH, Weinheim 2007, ISBN 978-3527-32141-4.

Schmidt, O. (ed.) et al.(2003): *Contributions to Juniperus communis.* Published by the Bavarian State Institute for Forestry and Forest Economics (LWF). LWF, 41, Freising 2003.

Schutt, P., Schuck, H. J., & Stimm, B. (2007). Encyclopaedia of forest botany. Morphology, pathology, ecology and systematics of important tree species. ecomed, Landsberg, new edition.

Schutt, P. and Schlimm, B. (2006): *Pinus sylvestris L.* Enzyklopadie der Holzgewachse (pp. 32pp), Edition 45 Erg. Lfg., ecomed-Verlag, Landsberg/Lech-Munchen.

Schweingruber FH (1996) Tree rings and environment. Dendroecology. Paul Haupt Verlag, Bern; 609 pages.

Schweingruber, FH. (1983): The tree ring. Location, methodology, time and climate in dendrochchronology. P. Haupt Verlag, Bern, Stuttgart.

Sheppard, P. R. (2010): Dendroclimatology: extracting climate from trees. *Wiley Interdisciplinary Reviews: Climate Change,* 1(3), 343-352, doi:10.1002/wcc.42.

Thomas P. A., El-Barghati M., Polwart A. (2007), Biological Flora of the British Isles: *Juniperus communis* L.. Journal of Ecology, 95: 1404-1440. doi: 10.1111/j.l365-2745.2007.01308.x.

Togneffi, R., Longobucco, A., Raschi, A.&Jones, M.B. (2001) Stem hydraulic properties and xylem vulnerability to embolism in three co-occurring Mediterranean shrubs at a natural CO2 spring. *AustralianJournalofPlantPhysiology,* 2 8 , 257-268.

Togneffi, R., Raschi, A. &Jones, M.B. (2002) Seasonal changes in tissue elasticity and water transport efficiency in three co-occurring Mediterranean shrubs under natural long-term CO2 enrichment. *Functional Plant Biology,* 2 9 , 1097-1106.

Wang, X., Yang, B., & Ljungqvist, F. C. (2019). The vulnerability of Qilian juniper to extreme drought events. *Frontiers in Plant science, 10,* 458686.

Weber, P., Bugmann, H., & Rigling, A. (2007). Radial growth responses to drought of *Pinus sylvestris* and Quercus pubescens in an inner-Alpine dry valley. *Journal of Vegetation Science, 18*(6), 777-792.

Wigley TML, Briffa KR, Jones PD (1984) On the Average Value of Correlated Time Series, with Applications in Dendroclimatology and Hydrometeorology. Journal of climate and Applied Meteorology, 23, 201-213.

Willson, C. J., & Jackson, R. B. (2006). Xylem cavitation caused by drought and freezing stress in four co-occurring Juniperus species. *Physiologia Plantarum, 127*(3), 374-382.

Wilson, R., Anchukaitis, K., Briffa, K. R., Buntgen, U., Cook, E., D'arrigo, R.,... & Zorita, E. (2016): Last millennium northern hemisphere summer temperatures from tree rings: Part I: The long term context. *Quaternary Science Reviews,* 134,1-18, doi:10.1016/j.quascirev.2015.12.005.

Wimmer, R., & Kriebitzsch, W. U. (2002). Scots pine (*Pinus sylvestris* L.) in Germany: Origin and utilisation. Allgemeine Forst- und Jagdzeitung, 173(1), 11-20.

7 Appendix

Table AI: Site and tree characteristics of the sampled Juniperus communis individuals (vitality classes: H = high (VK1), M = medium (VK2), N = low (VK3))

ContinuedNo.	Coordlnates	Umtang fcm]	DurctimcSSer (cm}	Vitality class	Floor HëПe (cm)	depth (cm)	Female
1	47=13'D.831 N 10=59 '14.295 E	21	6.7	H	220		7.5
1	47=13'0.854 N 10=59'14.298 E	25.5	9.1	H	310		12
2	47=13'0 .341 N 10=59 '14.355 E	20	6.4	H	330		4.5
4	47=13'1,515 N 10=59 '14,454 E	20	6.4	H	410		15
5	47=13'1,539 N 10=59 '14,374 E	19	5.1	H	210		13
6	47=13' 1,823 N 10=59 '14,115 E	173	5.5	H	310		10.5
7	47=13'1,475 N 10=59 '12,321E	15.5	43	H	320		10.5
8	47=13'2.013 bl 10=59'13.397 E	20	6.4	H	4C5		15
9	47=13'2.070 N 10=59 '13.393 E	31	9.8	H	3CD		15
1C	47=19'30.661 N 10=56'36.543 E	21.5	6.8	H	290		6
11	47=16'36.752 N 10=59'23.957 E	15	4.8	H	2CD		6
12	47=19'36.752 N 10=59'23.957 E	13	4.1	H	260		8
12	47=16'533 DON 10=59'12.434 E	20.5	6.5	M	40 0		14
14	47=13'2.301 U 10=59 9.769 E	21	6.7	H	330		14
15	47=13'1.101 N 10=59 '13.425 E	17.5	5.6	M	350		7.5
IE	47=13'2.86D bJ 10=59'12.045 E	33	10.5	H	43D		19
17	47=13'3.01014 10=59'11.333 E	12	3.8	H	20 D		B
IE	47=19'53.903 N 10=59'12.930 E	ia.5	53	H	280		9
IE	47=13'1,729 N 10=59 '10,956 E	12	3.8	M	26D		9
20	47=13'0.946 N 10=59'11.433 E	24	7.6	M	300		7.5
21	47=19'5922 53 Г4 10=59'10.274 E	10	3.2	H	20 0		16
22	47=13'0.496 N 10=59 '12.012 E	19	5.7	H	205		13
22	47=19' 593 D6N 10=59'10.332 E	16.5	5.3	H	30 D		14
24	47=13'D.231 bJ 10=53'12.943 E	24.5	7.8	H	350		13
2;	47=13' D .469 N 10=59 '11.994 E	19	5.7	H	20 D		14 W
2E	47=13'59319 N 10=59'12.435 E	13	4.1	H	230		8.5
2/	47=19'53.994 N 10=59'12.421 E	15	4.8	N	20 D		10 w
2£	47=16'53315 N 10=59'12.477 E	14	4.5	H	20 0		7
2E	47=13'0.231 N 10=59'12.943 E	15	4.8	H	205		5
30	47-18'0.661 H 10-58'12.975 E	21	6.7	N	225		6
31	47"1S'D.632H 10-58'13.181 E	15	4.8	H	300		7
32	47°1S' 0.477 H 10-58 '13.678 E	23	7.3	H	305		9
33	47°18'588B5 N 10=58'11.376 E 47°1B'58815N ИГ5В'12.477	22.5	7.2	M	230		7
34	E	16	5.1	M	260		12.5
35	47°18'58.D72 N 10=58'12.054 E	16.5	5.3	M	320		Э.5
36	47°1B'58.D37 N 10=58'13.452 E	24.5	7.8	H	350		9
37	47-18'58.341 N 10=58'13.227 E	15	4.8	M	240		10
38	47°18'58887N 10=58'13.954 E	16.5	5.3	M	3CD		17
	47-18'58.318 N 10=58'13.754						

No.	Coordinates					
39	E	19.5	6.2	M	500	11.5
40	47-18'58.747 N 10=58'14.472 E	15	4.8	N	480	10
41	47-18'58.234 N 10=58'14.907 E	21	6.7	H	23D	10
42	47-18'58.733 N 10=58'15.109 E	22	7.0	H	320	Э.5
43	47-18'58.670 N 10=58' 15.223 E	33	10.5	M	320	11 W
44	47-18'58.648 N 10=58'14.991 E	12.5	4.0	N	25D	10.5 W
45	47-18'0.238 H 10-58'14.507 E	11	3.5	M	200	7.5 W
46	47-18'1.132 H 10-58'15.010 E	11	3.5	H	21D	7
47	47-18'0.733 H 10=53'15.215 E	12	3.8	M	260	11
48	47-18'1.621 H 10-58'12.316 E	21	6.7	M	450	7.2
49	47-18'1.230 H 10-58'12.781 E	15.5	4.3	M	215	5
50	47°1S'1.6K H 10'58'11.336 E	15	4.8	M	210	7 W
51	47-18'2.416 H 10-5B'12.162 E	30	8.6	H	320	3
52	47'18'2.371 H 10'58'0.831 E	13	4.1	M	300	9
53	47-18'2.148 H 10-58'11.577 E	16	5.1	M	25D	3
54	47-18'57.258 N 10=58'13.731 E	17	5.4	M	320	7
55	47-18'57.358 N 10=58'13.883 E	15.5	4.3	M	350	4
56	47-18'57.185 N 10=58'14.105 E	20	6.4	M	330	9
57	47-18'57.832 N 10=5B'U.103 E	14	4.5	N	36D	9
58	47-18'56.882 N 10=58'14.023 E	15.5	4.3	M	235	10 W
59	47-18'57.746N 10=58'15.101 E	11	3.5	N	275	3
60	47-18'32.855 N 10=58'6.386 E	14	4.5	H	210	6.3
61	47-18'55.743 N 10=58'8.357 E	13	4.1	M	2CD	6.2
62	47"1B' 55.765 N 10-58'8.329 E	18	5.7	H	400	9.2
63	47°1B'55J16 N 10-58'9.920 E	23	7.3	H	420	8.9
64	47°1B'56830 N 10' '5B'9.222 E	13	4.1	M	210	3
65	47°18'56853 N 10-58'9.299 E	28	8.9	H	520	4.2 W
66	47°1B'57.OD3 N 10-58'9.462 E	19	6.1	M	420	6.2
67	47*18'56866 N 10-58'9.451 E	15	4.8	N	310	5.3
63	47°1E'55893 N 10-58'9.955 E	11	3.5	N	235	6.7
69	47-IB'57.619 N 10-58'11.009 E	13	4.1	H	380	9.5
70	47-18' 58.455 N 10-58'11.601 E	7	2.2	N	200	9.6
71	47-IB'58.127 N 10-58' 10.346 E	10	3.2	N	230	16
72	47-IB' 57.662 N 10-58'9.715 E	IB	5.7	N	200	9.5 W
73	47-IB[1] 57.347 N 10' '5B7.4D0 E	19	6.1	H	430	11
74	47°1B'56^55 N 10-58'6.228 E	11	3.5	N	200	9
75	47-18' 57.352 N 10-58'6.621 E	12	3.8	M	200	9

76	47-18'55.558 N 10-58'6.561 E	17	5.4 N	450	9.5
77	47-18' 55.767 N 10-58'5.448 E	13	**4.1** N	330	5
73	47-18'55865 N 10-58'5.635 E	15	4.8 N	400	5.2
79	47-18'55 J73N 10-58'5.656 E	11	3.5 N	430	5
80	47-18' 56.622 N 10-58'5.079 E	12	3.8 H	510	6.5
81	47-18'56.151 N 10-58'5.606 E	17	5.4 N	310	7
82	47-18'57.011 N 10-58'5.223 E	15	4.8 N	320	5.6
83	47-18' 57.005 N 10-58'5.223 E	11	3.5 N	230	5.2
84	47-18'57,000 N 10-58'5,228 E	16	5.1 N	350	6
85	47-18'56.654 N 10-58'4.914 E	10	3.2 N	250	5.4
86	47°1E'56.1B4 N 10-58'5.668 E	26	8.3 H	450	7.3
87	47-18'56841 N 10-58'6.226 E	20	6.4 N	230	6.7
88	47-18'56.643 N 10-58'6.122 E	14	4.5 N	250	6.9
89	47-18'56.551 N 1O°5B'6.211 E	12	3.8 N	210	6.5
90	47-lB' 56.5S 1 N 10-58'6.618 E	11	3.5 M	520	5
91	47-lB'56820 N 10-58'6.039 E	12	3.8 N	230	5.4
92	47-lB'57.625 N 10-58'6.043 E	17	5.4 M	430	7.3
93	47-lB' 57.538 N 10-58'5.326 E	10	3.2 M	530	9.1
94	47-lB' 57.5B8N 10-58'5.723 E	16	5.1 N	510	6.2
95	47-18' 58845 N 10-58'6.137 E	14	4.5 N	470	6.1
96	47-lB' 57.728 N 10-58'6.389 E	11	3.5 N	430	7.3
97	47-lB' 57.730 N 10-58'6.396 E	9	2.9 N	410	5.7
98	47°1B'57816 N 10-58'6.620 E	17	5.4 M	490	6.1
99	47-18' 57.138 N 10-58'7.224 E	10	3.2 N	250	11.2
100	47-18' 57864 N 10-58'6.661 E	9	2.9 N	250	7.3
101	47-lB' 58.071 N 10-587.767 E	9	2.9 M	240	6.7
102	47°1B' 58.072 N 10-587.767 E	24	7.6 H	330	7.2
103	47°1B' 58.560 N 10-587.320 E	21	6.7 H	510	10.2 W
104	47°1B'588D2 N 10-58'9.679 E	19	6.1 H	630	9.3
105	47-lB'58.403 N 10-58'9.172 E	12	3.8 H	380	10.3
106	47-18'588 75 N 10-58'9.022 E	2D	6.4 H	620	6.3
107	47=19'58.754 N 1(F59'3.4D7 E	10	3.2 M	280	5.4
108	47=19'58.756 N 1(F59'3.4D8 E	14.5	4.6 M	320	5.9
109	47=19' 58.9 54 N 1(F59'3.435 E	22.5	7.2 M	610	10.1 W
110	47=19'58366 N 1(F59'9.318 E	19.5	6.2 N	590	8.9 W
111	47=13'D.4C2N 10=583.683 E	24.5	7.8 **M**	780	7.8
112	47=18' 59.58 **1** N 1(F58' 1C .849 E	20	6.4 H	520	3.8
113	47=13'0.896 **N** 10-59'10.128 E	21	6.7 H	280	5.8 W
114	47=13'1.225 N 10=587.785 **E**	16	5.1 H	220	10.2 W
115	47=13'D.924 N 10-58'8.686 E	17	5.4 N	210	6.3
116	47=13'1.147 N 10=59'9.172 E	29	9.9 H	720	7.8 W
117	47=13'1.136 **N** 10=59'7.982 E	17	5.4 H	690	8.2 W
119	47=13'0.487 N 10=59'9.433 E	23	7.3 M	380	10.8
119	47=13'D.**534 N** 10=583.293 **E**	19.5	6.2 H	360	6.3
12D	47=13'1.289 N 10=593.585 **E**	23	7.3 H	320	5.2 W

Pines	Coordinates	Diameter Circumference (cm)	(cm)	Vitality class	High (cm)	Floor depth
1	47°19'0.664 N 10°58'9.075 E	131	41.7	M	12.3	10.2
2	47°19'0.943 N 10°58'9.585 E	101	32.2	M	10.8	9.8
3	47°19'0.369 N 10°58'10.064 E	96	30.6	N	9.8	8.9
4	47°19'0.797 N 10°58'11.291 E	82.5	26.3	M	12.3	10.1
5	47°19'1.510 N 10°58'10.176 E	66	21.0	M	9.6	11.3
6	47°19'1.514 N 10°58'11.079 E	38.5	12.3	N	9.2	7.6
7	47°19'1.215 N 10°58'11.852 E	74	23.6	H	7.6	13.5
8	47°18'58.357 N 10°58'10.796 E	93.5	29.8	H	10.6	12.3
9	47°18'59.263 N 10°58'10.047 E	68	21.7	H	11.8	10.6
10	47°18'59.700 N 10°58'13.288 E	84	26.8	H	12.3	11.2
11	47°18'59.380 N 10°58'13.027 E	58	18.5	H	10.2	10.8
12	47°18'59.475 N 10°58'12.388 E	60	19.1	N	9.2	11.3
13	47°18'59.720 N 10°58'14.820 E	50	15.9	N		1012.1
14	47°18'59.261 N 10°58'13.294 E	40	12.7	H	10.5	13.1
15	47°18'59.271 N 10°58'13.342 E	81.5	26.0	H	10.2	13.5

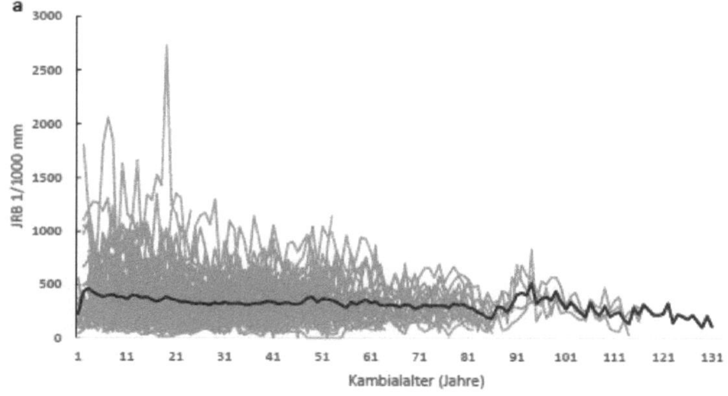

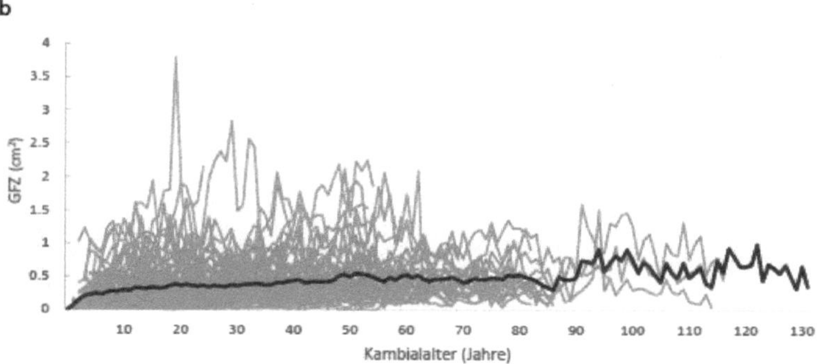

FigureA1a-b: Tree-ring width chronologies (JRB) and GFZ of all dated Juniperus communis individuals (n=80) and their mean chronology (red line). JRB chronologies by cambial age.

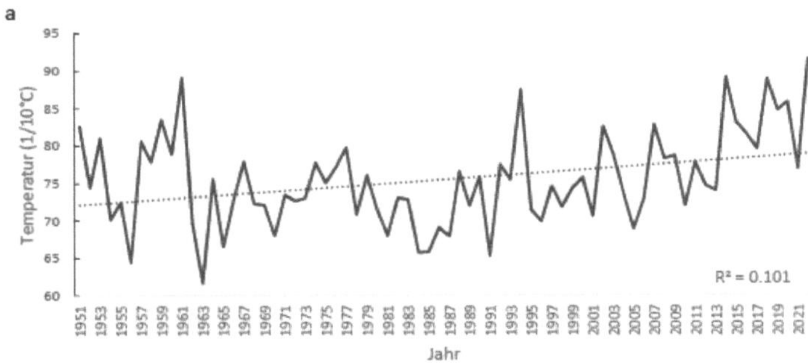

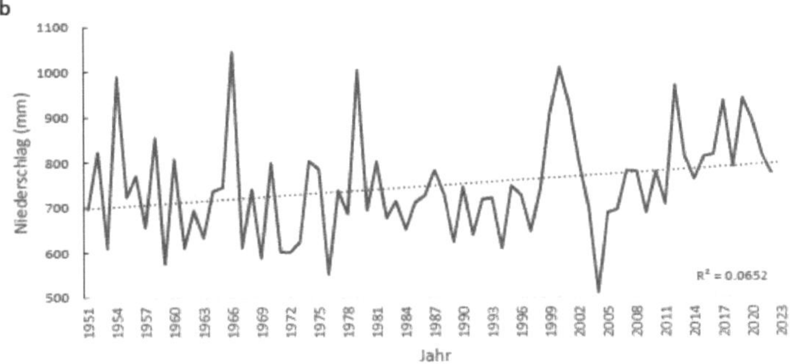

Figure A2: Annual precipitation and temperature curves over the period from 1951 to 2023

Figure A3 shows the precipitation from the previous year and the dry year for each season and once as an annual average. What can be clearly seen is that the summer precipitation (yellow) always drops significantly in the previous year and the spring (green), autumn (red) and winter precipitation (blue) are significantly lower in the dry year and not in the previous year. The annual precipitation (black) only drops significantly in 2004. Shown here is the deviation of precipitation from normal precipitation.

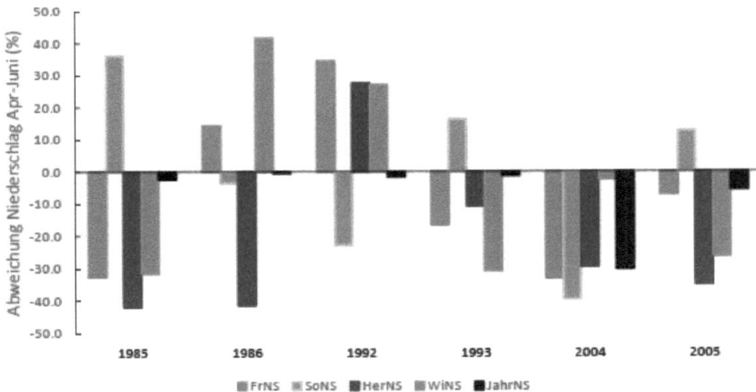

FigureA3: Deviation of seasonal precipitation from the long-term average in the extreme growth years 1986, 1993 and 2005 and their previous years
FrNS: Precipitation March-May; SoNS: Precipitation June-August; HerNS: Precipitation September-November; WINS: Precipitation December-February; JahrNS: Precipitation January-December

Figure A4: Deviation of precipitation from April to June from the long-term average in the extreme growth years 1986, 1993 and 2005 and their previous years.

Figure A5a-c shows the stress indices (resistance, recovery and resilience) for UKs 1-3 for the extreme growth years 1986, 1993 and 2005.

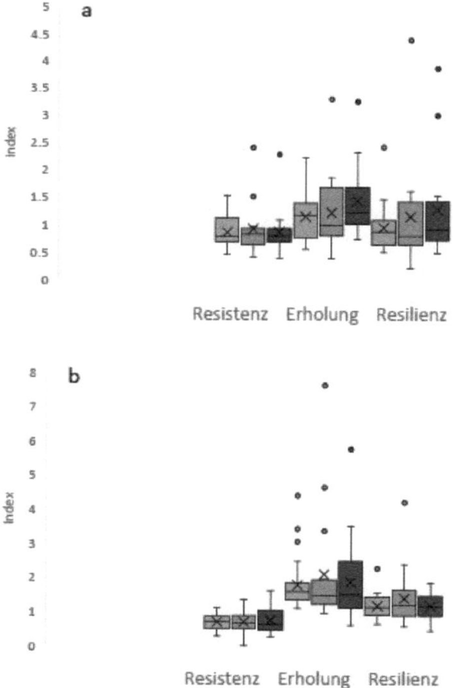

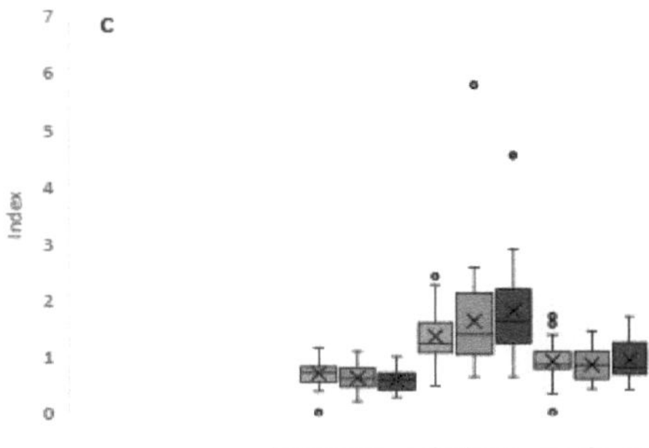

Figure A5: Scattering of the stress indices in the vitality classes (I-3) for the extreme growth years 1986, 1993 and 2005
Boxplot represents the resistance, recovery and resilience of the individual VCs (green: VC 1; orange: VC 2; red: VC 3). The
dots represent the highest or lowest value. (Whiskers = standard deviation Line in boxplot = median value x = mean value).
(a)   Stress indices of the three vitality classes for the year 1986
(b)   Stress indices of the three vitality classes for the year 1993
(c)   Stress indices of the three vitality classes for the year 2005

In Figure A9, the *Juniperus* communis *individuals* were divided into < 50 and > 50 years and
the stress indices are shown for the extreme growth years 1986, 1993 and 2005. In 1986 and
2005, the < 50 year old individuals show a higher stress index than the older specimens. In
1993, the recovery of the > 50 year old individuals is better than that of the younger ones,
but the resistance and resilience is better in the younger individuals.

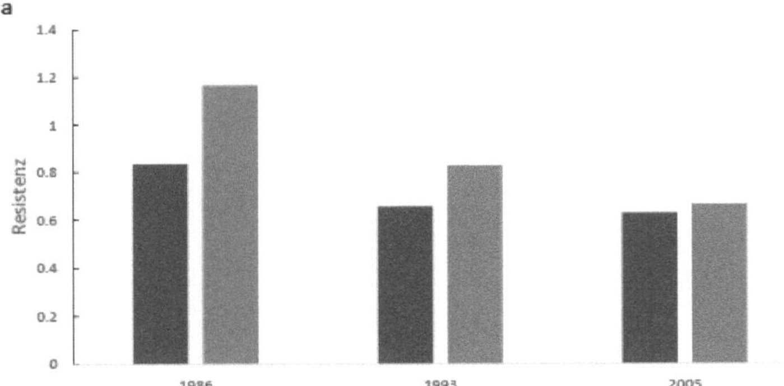

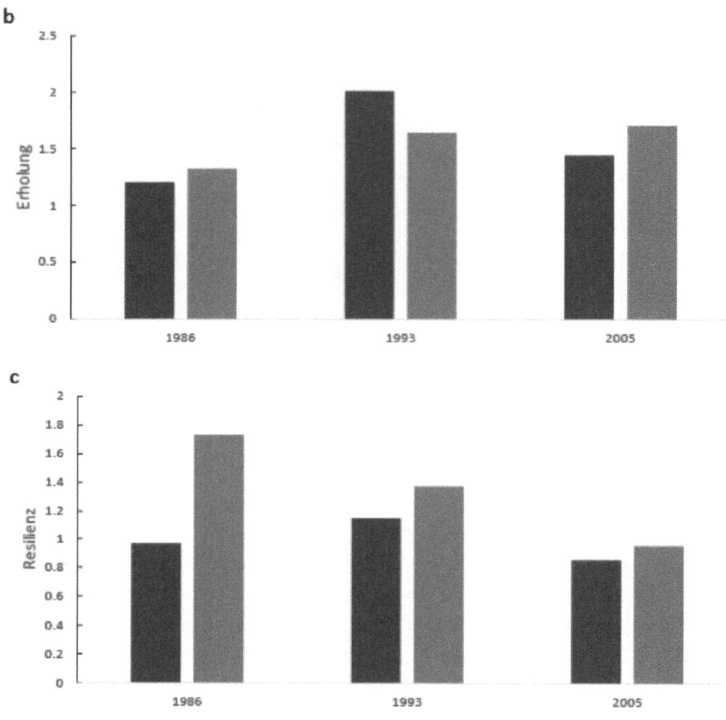

Figure A9a-c: Stress indices (mean value; resistance, recovery and resilience) for the extreme growth years in the age groups < 50 and > 50 years (blue bars: > 50 years; n: 45; orange bars < 50 years; n: 35)

Printed by Books on Demand GmbH, Norderstedt / Germany